U.S. Military Operations in Afghanistan

Effectiveness of Psychological Operations 2001–2010

Arturo Munoz

Prepared for the Marine Corps Intelligence Activity
Approved for public release; distribution unlimited

NATIONAL DEFENSE RESEARCH INSTITUTE

The research described in this report was prepared for the Marine Corps Intelligence Activity. The research was conducted within the RAND National Defense Research Institute, a federally funded research and development center sponsored by the Office of the Secretary of Defense, the Joint Staff, the Unified Combatant Commands, the Navy, the Marine Corps, the defense agencies, and the defense Intelligence Community under Contract W74V8H-06-C-0002.

Library of Congress Cataloging-in-Publication Data

Munoz, Arturo, 1949-
U.S. military information operations in Afghanistan : effectiveness of psychological operations 2001-2010 / Arturo Munoz.
   p. cm.
 Includes bibliographical references.
 ISBN 978-0-8330-5151-6 (pbk. : alk. paper)
 1. Afghan War, 2001---Psychological aspects. 2. Afghan War, 2001---Propaganda. 3. United States—Armed Forces—Afghanistan. 4. Psychological warfare—History—21st century. I. Title.

   DS371.412.U16 2012
   958.104'78—dc23

                                                                    2011049898

Cover photo by Staff Sgt. Samuel Bendet, U.S. Air Force.

Published 2012 by the RAND Corporation
1776 Main Street, P.O. Box 2138, Santa Monica, CA 90407-2138
1200 South Hayes Street, Arlington, VA 22202-5050
4570 Fifth Avenue, Suite 600, Pittsburgh, PA 15213-2665
RAND URL: http://www.rand.org/
To order RAND documents or to obtain additional information, contact
Distribution Services: Telephone: (310) 451-7002;
Fax: (310) 451-6915; Email: order@rand.org

Preface

The U.S. Marine Corps (USMC), which has long recognized the importance of influencing the civilian population in a counterinsurgency (COIN) environment, asked the RAND National Defense Research Institute to evaluate the effectiveness of U.S. military (USMIL) information operations (IO) and psychological operations (PSYOP) in Afghanistan from 2001 to 2010 based on how well messages and themes are tailored to target audiences. This monograph responds to that request. It should be emphasized that this report does not cover the significant changes in IO and PSYOP definitions, doctrine, organization, and implementation in the field that have taken place since 2010. When the text refers to the present, or the current situation, it generally means 2010.

This research was sponsored by the Marine Corps Intelligence Activity (MCIA) and conducted within the Intelligence Policy Center of the RAND National Defense Research Institute, a federally funded research and development center sponsored by the Office of the Secretary of Defense, the Joint Staff, the Unified Combatant Commands, the Navy, the Marine Corps, the defense agencies, and the defense Intelligence Community.

For more information on the RAND Intelligence Policy Center, see http://www.rand.org/nsrd/ndri/centers/intel.html or contact the director (contact information is provided on the web page).

Contents

Figures

Tables

Summary

Background

The United States has been engaged in conflict in Afghanistan for nearly a decade. From the outset, U.S. leaders recognized the importance of winning the support of the Afghan population, given the country's history of antipathy toward foreign armies. Initial efforts to influence the population met with some success, but ensuing years have seen rising disenchantment with the Hamid Karzai administration and coalition forces. The USMC is heavily engaged in Afghanistan, primarily in Helmand province, where the Taliban had controlled extensive areas prior to the U.S. offensive and remain a potent force. The honing of messages to sway the population is critical to the ongoing campaign to establish permanent Government of the Islamic Republic of Afghanistan (GIRoA) control over the province. Accordingly, in a lessons-learned context, the MCIA requested an assessment of the effectiveness of prior efforts so that it could improve its own operations in this area. It should be noted that the research focused exclusively on the U.S. military. Operations conducted by North Atlantic Treaty Organization (NATO) forces in Afghanistan were not reviewed.

Purpose

This monograph reviews the effectiveness of USMIL IO and PSYOP in Afghanistan from late 2001 to 2010. The other four core capabilities employed by IO to achieve desired combatant commander effects—

electronic warfare (EW), military deception (MILDEC), computer network operations (CNO), and operations security (OPSEC)—are not covered. It should be noted that, in June 2010, the U.S. Department of Defense (DoD) officially replaced the term *PSYOP* with *military information support operations (MISO)*.[1] This monograph continues to use the term *PSYOP*, however, because that was the term in force during the research and it is unclear at the time of this writing what will change and what will remain the same under the new term. The basic goal of this monograph is to summarize the diverse PSYOP initiatives undertaken, evaluate their effectiveness, identify strengths and weaknesses, and describe the way forward, including making certain specific recommendations for improvements. Special attention was paid to how well PSYOP initiatives were tailored to target audiences, primarily the Pashtuns who are the dominant population in the conflictive areas and the main support of the Taliban insurgency.

Although this monograph focuses mainly on the effectiveness of themes and messages among Afghan target audiences, it also discusses IO and PSYOP doctrine and organization because of their impact on the effectiveness of messaging. Most operations conducted in Afghanistan under the rubric of IO pertain to the PSYOP core capability, but IO practitioners implementing these operations often view PSYOP as a separate activity. DoD was aware of the confusion over terminology and repeatedly issued guidance seeking to clarify differences between these "overlapping but distinct concepts."[2] Nonetheless, as has been noted in previous studies, *IO* has become a substitute term for *PSYOP*.[3] Few practitioners in the field seemed to follow the strict interpretation

---

[1]   Alfred Paddock, Jr., "PSYOP: On a Complete Change in Organization, Practice, and Doctrine," *Small Wars Journal*, June 26, 2010.

[2]   U.S. Department of Defense, *Consolidated Report on Strategic Communication and Information Operations*, submitted to Congress, March 2010a, p. 5.

[3]   See Todd C. Helmus, Christopher Paul, and Russell W. Glenn, *Enlisting Madison Avenue: The Marketing Approach to Earning Popular Support in Theaters of Operation*, Santa Monica, Calif.: RAND Corporation, MG-607-JFCOM, 2007.

of IO as solely a coordinating or integrating function.[4] This monograph contains various direct quotes from interviews, briefing slides, and other written material that refers to IO products, IO messages, and IO campaigns, but the official point of view insists that, doctrinally, IO are strictly a coordinating or integrating function that should not produce specific products.

For their part, PSYOP specialists tended to view the emphasis on IO as disproportionate and their own role as undervalued. They felt marginalized even before the term *PSYOP* was abolished officially in 2010 due to its perceived negative connotations. Whereas the Taliban implemented a unified anti-U.S. propaganda campaign, the United States subdivided its counterpropaganda capabilities, creating separate entities with overlapping missions and definitions. This could have negative ramifications for the overall effort to create a skilled cadre of specialists to deal with this nonkinetic aspect of asymmetrical warfare. Although the existing division of labor between IO, PSYOP, and strategic communication makes sense on a theoretical level, in practice, in the Afghan theater during the period in question, it did not seem to be the most-efficient way to marshal limited resources against the enemy's relentless propaganda offensive.

Findings

As a U.S. general asked in a 2010 Kabul meeting that I attended, the key question is, "Are we losing the information war?" The overall response is neither affirmative nor negative. This monograph contains reports of specific operations that were very successful in achieving IO objectives. However, there are also examples of operations that did not resonate with target audiences and even some that had counterproductive effects. If the overall IO mission in Afghanistan is defined as convincing most residents of contested areas to side decisively with the

---

[4] An example of the field perspective can be seen in the excellent article by Ensign Robert J. Bebber, "Developing an IO Environmental Assessment in Khost Province: Information Operations at PRT Khost in 2008," *Small Wars Journal*, February 28, 2009.

Afghan government and its foreign allies against the Taliban insurgency, this has not been achieved. Even when USMIL IO and PSYOP take all the right steps, message credibility can be undercut by concern among Afghans in contested areas that their own government, widely perceived as weak and corrupt, will not be able to protect them from vengeful Taliban once U.S. and NATO forces withdraw. Civic action and development projects are greatly appreciated, but some public-opinion surveys suggest that both the Taliban and U.S. and NATO forces are viewed negatively.

The biggest PSYOP successes have been in the area of face-to-face communication and the new emphasis on meetings with jirgas (local councils of elders), key-leader engagements, and establishing individual relationships with members of the Afghan media. Also, the concept of every infantryman being a PSYOP officer, as carried out by the 1st Battalion (Bn) 5th Marines and other units, is also very effective. In this respect, the success of civic action and development projects in promoting a positive image of the U.S. military and the Afghan government should be pointed out, although this varies greatly among localities.

On the negative side of the ledger, the most-notable shortcoming has been the inability to effectively counter the Taliban propaganda campaign against U.S. and NATO forces on the theme of civilian casualties, both domestically and internationally. Nonetheless, it should be stressed that this Taliban propaganda success does not translate into widespread popular support for the Taliban movement. On the contrary, most polls indicate that the great majority view the Taliban negatively, which suggests that their messaging has not achieved all of its objectives either. Although results of district-level polling vary greatly, the Taliban overall do not seem to enjoy great popularity. PSYOP products highlighting specific acts of Taliban terrorism, such as destruction of schools and the killing of schoolteachers, do discredit the insurgency. Nonetheless, throughout 2001–2010, audiences generally have not responded to offers of rewards for information on terrorist leaders.

PSYOP themes and messages tend to be more effective when they reflect Afghans' yearning for peace and progress. [5] It should be stressed at this point, moreover, that the Afghan audience is not homogenous. On the contrary, Afghan society is deeply divided by ethnic, tribal, and regional cleavages, and this affects PSYOP target audience selection and analysis. The key audience for counterinsurgency objectives is the Pashtuns, who make up about 42 percent of the national population and inhabit those areas where the Taliban insurgency is strongest. Failure to adequately incorporate Pashtun perceptions and attitudes can negate the potential effectiveness of many PSYOP products. In this respect, USMIL PSYOP have been criticized for not adequately countering the Taliban's manipulation of Pashtun religious and nationalistic sentiments. Also, there has been variation over time. Such themes as the promotion of democracy and participation in elections seemed to have better audience reception during 2001–2005 than they had in later years, including the most-recent elections, in 2009 and 2010. The reason for this decline in effectiveness has less to do with the content of the products than with the growing disillusionment over the regime's corruption and its inability to provide security and services. Moreover, credibility of USMIL IO and PSYOP messaging is undercut by contradictory public statements made by GIRoA and the International Security Assistance Force (ISAF) or U.S. government spokespeople regarding air strikes, collateral damage, night raids, and electoral fraud.

This underscores the notion that external factors over which PSYOP personnel have no control could ultimately determine the acceptance of their messages among target audiences.

Table S.1 contains the assessment of nine major themes of PSYOP efforts, rating them as effective, mixed, or ineffective.

---

[5] The best evidence of this yearning is in the numerous public-opinion polls conducted in Afghanistan during the period. Also, interviews with IO and PSYOP officers returning from the field often indicate that the jirgas express very concrete ideas of desired peace and progress. This author has observed target audiences' positive reactions to the peace and progress theme, and this is well articulated in CDR Larry LeGree, U.S. Navy, "Thoughts on the Battle for the Minds: IO and COIN in the Pashtun Belt," *Military Review*, September–October 2010, pp. 21–32

Table S.1
Assessment of Major Themes in Psychological Operations

Theme	Assessment		
	Effective	Mixed	Ineffective
The war on terror justifies U.S. intervention.			Ineffective
Coalition forces bring peace and progress.	Effective 2001–2005	Mixed 2006– 2010	
Al-Qai'da and the Taliban are enemies of the Afghan people.		Mixed	
Monetary rewards are offered for the capture of al-Qai'da and Taliban leaders.			Ineffective
Monetary rewards are offered for turning in weapons.		Mixed	
Support of local Afghans is needed to eliminate IEDs.		Mixed	
U.S. forces have overwhelming technological superiority over the Taliban.	Effective 2001–2005	Mixed 2006– 2010	
GIRoA and ANSF bring peace and progress.		Mixed	
Democracy benefits Afghanistan, and all Afghans need to participate in elections.	Effective 2001–2005	Mixed 2006– 2010	

NOTE: IED = improvised explosive device. ANSF = Afghan National Security Forces.

Interviews with IO and PSYOP personnel who have served in Afghanistan, which have been corroborated by various other studies, point to several organizational problems impeding effectiveness of their mission. These include inadequate coordination between IO and PSYOP, long response times in the approval process, lack of IO and PSYOP integration in operational planning, lack of measures of effectiveness (MOEs), and an inability to exploit the informal, oral Afghan communication tradition. However, these problems are not universal. Some commanders, for example, have become well known in theater for their insistence on integrating IO and PSYOP with their operational planning, and there are cases in which counterpropaganda responses to Taliban charges against U.S. forces have been rapid and well conceived.

Recommendations

To improve the effectiveness of IO and PSYOP, this monograph makes the following recommendations.

Hold a conference of IO and PSYOP personnel who have served in Afghanistan to define best practices. The objective would be to define best practices based on their recent experiences in the field and make specific recommendations for operational, organizational, and doctrinal reforms.

Use local focus groups to pretest messages. Failure to take into account cultural, social, political, and religious factors is a major deficiency in PSYOP campaigns. Using focus groups to pretest messages can help correct this deficiency, but the focus groups' membership must closely parallel that of the target audience.

Conduct public-opinion surveys for target-audience analysis and posttesting. Considerable polling and interviewing are being conducted in Afghanistan, some of it USMIL sponsored. Significant work on human terrain mapping and cultural intelligence has also been accomplished. However, much-better use of these data could be made to develop PSYOP themes and messages. The surveys should be keyed to specific PSYOP campaigns. Moreover, the emphasis should be on district-level polling, as opposed to national-level polls, which might not be representative of target audiences in conflictive areas. Survey research can provide quantitative baselines and trend analyses of key attitudes held by the target audience. In addition, it can help predict attitude change based on knowledge of underlying attitude structures and, thereby, help develop appropriately targeted messages. Also, polling can be effective in posttesting specific PSYOP products, helping to determine whether the audience reacts as intended.

Use key communicators to help develop and disseminate messages. Messages are more credible if they come from a figure who already enjoys prestige within the target audience and is already considered a credible source of advice and information. In Afghanistan, key communicators can vary greatly between communities. A key communicator could be an Islamic cleric, a traditional chief, an educated schoolteacher, a wealthy merchant known for providing charity, a local leader who maintains a loyal following, or a government official, among others. Moreover, in the Pashtun tribal context, a key communicator might not necessarily be an individual but could be a collective group, such as the elders comprising a jirga. This monograph proposes that the traditional PSYOP role of the key communicator be expanded. Key communicators should be considered partners in developing messages, contributing not only to the wording but also to the content.

Harmonize IO doctrine and practice, and implement greater integration with PSYOP and public affairs (PA). The current disconnect between official IO doctrine and how it is practiced in the field is counterproductive. The situation has been further complicated by the recent elimination of the term PSYOP, entailing, in the words of U.S. Special Operations Command (USSOCOM) commander ADM Eric T. Olson, a "complete change in organization, practice, and doctrine."[6] That being the case, at the time of this writing, clarification of the revised PSYOP mission is needed. Also, the current division between PSYOP and PA works to the advantage of Taliban propagandists, who routinely accuse U.S. forces of needlessly causing civilian casualties. Closer coordination between PSYOP and PA would enhance counterpropaganda effectiveness.

---

[6]   See Paddock, 2010.

Acknowledgments

I would like to highlight the special contributions to this study made by RAND analyst Sara Beth Elson. RAND research assistant Jonathan Vaccaro, who recently returned from a combat tour in Afghanistan, provided excellent insights from his personal observations conducting IO, as well as timely research on the subject matter. Wali Shaaker, a native Pashtun who has assisted U.S. forces in planning and implementing PSYOP campaigns in Afghanistan, afforded very good observations and recommendations. Christopher Paul, author of various works on IO, including a reference handbook, wrote an extremely useful critique, which helped achieve sharper focus on key points. He provided texts and references, which were integrated into the text, as were the references and insightful critique provided by Mir Sadat of the National Defense Intelligence College. Also highly useful were the graphs prepared by research assistant Joya Laha illustrating findings of Afghan public-opinion polls, as well as main PSYOP themes and messages.

In Kabul, the contributions of LTC John "Chip" Bircher, special assistant to the director of communication at the International Security Assistance Force (ISAF) Headquarters (HQS), and Col Jeffrey L. Scott, fusion cell chief, Communication Directorate, ISAF HQS, were invaluable. Extensive IO and PSYOP materials they provided have been incorporated into this study. I would also like to express great appreciation for the information provided by LTC Maria Metcalf, strategic planner for the director of communication, ISAF, and Eric Sutphin in Target Audience Analysis, Combined Joint Psychological Operations

Task Force (CJPOTF), ISAF. At USSOCOM J-239, SOCOM Support Team Texas, I offer special thanks to LTC Kurt Saffer and Steve Kite for taking the time to review this monograph in its early stages and offer comments.

The comments and advice provided by Austin Branch, senior advisor for strategy and plans at the Office of the Under Secretary of Defense for Intelligence, Information Operations and Strategic Studies, have also been very helpful. Douglas Friedly, a senior analyst at the Office of the Under Secretary of Defense for Policy, IO Directorate, took the time to do a page-by-page review with an emphasis on key doctrinal issues, which was extremely valuable. Likewise, Eric V. Larson of RAND provided a careful review, which is greatly appreciated. Finally, I would like to give special recognition and thanks to the editors of the *psywarrior.com* website for their permission to reproduce copies of various leaflets and posters disseminated in Afghanistan by the U.S. military.

Abbreviations

ANP	Afghan National Police
ANSF	Afghan National Security Forces
AOR	area of responsibility
Bn	battalion
BUB	battle update brief
CA	civil affairs
CAC	Combined Arms Center
CERP	Commanders' Emergency Response Program
CFC-A	Combined Forces Command–Afghanistan
CFR	Council on Foreign Relations
CIA	Central Intelligence Agency
CIVCAS	civilian casualties
CJPOTF	Combined Joint Psychological Operations Task Force
CJTF-76	Combined Joint Task Force 76
CNO	computer network operations
COIN	counterinsurgency

DoD	U.S. Department of Defense
EW	electronic warfare
FM	field manual
FOB	forward operating base
FY	fiscal year
G-2	Army or Marine Corps component intelligence staff officer
GIRoA	Government of the Islamic Republic of Afghanistan
GWOT	global war on terrorism
HHQ	higher headquarters
HN	host nation
HQS	headquarters
ICOS	International Council on Security and Development
IED	improvised explosive device
IO	information operations
IOPB	information operations preparation of the battlefield
ISAF	International Security Assistance Force
J-239	U.S. Special Operations Command Support Team
JFC	joint force commander
JP	joint publication
JSOA	joint special operations area

MCIA	Marine Corps Intelligence Activity
MEDCAP	medical civic action program
MILDEC	military deception
MISO	military information support operations
MIST	military information support team
MOE	measure of effectiveness
MOP	measure of performance
MOS	military occupational specialty
NATO	North Atlantic Treaty Organization
NDAA	National Defense Aurhotization Act
NGO	nongovernmental organization
OEF	Operation Enduring Freedom
OIF	Operation Iraqi Freedom
OPCEN	operations center
OPLAN	operation plan
OPSEC	operations security
PA	public affairs
PAO	public affairs office
POTF	U.S. Psychological Operations Task Force
PRT	provincial reconstruction team
PSYACT	psychological action
PSYOP	psychological operations
PVO	private voluntary organization
RC	regional command

RCIED	radio-controlled improvised explosive device
RCU	rich contextual understanding
TF	task force
UNAMA	United Nations Assistance Mission in Afghanistan
USAID	U.S. Agency for International Development
USAIOP	U.S. Army Information Operations Proponent
USDA	U.S. Department of Agriculture
USMC	U.S. Marine Corps
USMIL	U.S. military
USSOCOM	U.S. Special Operations Command
UXO	unexploded ordnance

Introduction: Definition and Objectives of Psychological Operations in Afghanistan

Background

From the beginning of the U.S. military intervention in Afghanistan in 2001, psychological operations (PSYOP) were employed to gain popular acceptance for the overthrow of the Islamic Emirate, the presence of foreign troops, and the creation of a democratic, national government. During the initial period of this nation-building effort, it seemed that success was being achieved. However, disenchantment with the Karzai administration began to grow, augmented by increasing resentment against North Atlantic Treaty Organization (NATO) and U.S. military tactics negatively affecting local populations. The Taliban movement began to revive. Meanwhile, according to a 2009 paper written by COL Francis Scott Main, PSYOP capabilities in Afghanistan during that period declined:

> Since the U.S invasion of Iraq in 2003, the capability of the US PSYOP Task Force (POTF) in Afghanistan to disseminate strategic communications through regional media . . . steadily decreased. . . . PSYOP personnel and resources in Afghanistan were reduced in order to meet requirements for PSYOP operations [sic] in Iraq. . . . The ability of US forces to react to a nimble adversary [that] does not follow the same rules as the US is inadequate in Afghanistan. . . . The POTF . . . had suffered a steady decline of key personnel . . . from 2003 to 2007. . . . In April of 2006, the reassignment of all PSYOP Regional Battalion assets from Afghanistan left the POTF with only a tactical PSYOP

product development capacity . . . reducing the number of US soldiers developing PSYOP products from twenty-five to eight.[1]

It should be noted that the 2003 U.S. Department of Defense (DoD) *Information Operations Roadmap* went beyond lamenting the decline of PSYOP in Afghanistan, stating that this function was weak in general:

> We must improve PSYOP. Military forces must be better prepared to use PSYOP in support of military operations and the themes and messages employed in a PSYOP campaign must be consistent with the broader national security objectives and national-level themes and messages. Currently, however, our PSYOP campaigns are often reactive and not well organized for maximum impact.[2]

By the time GEN Stanley A. McChrystal arrived in July 2009 as commander of the International Security Assistance Force (ISAF), resurgent Taliban guerrillas were expanding their control over rural areas and stepping up terrorist attacks on government officials and supporters. Public-opinion polls showed a drastic decline in the image of the Afghan government, as well as U.S. and coalition military forces. Reversing this loss of popular support became critical, and this task fell within the purview of information operations (IO). General McChrystal undertook a thorough review of the situation and concluded,

---

[1]   COL Francis Scott Main, U.S. Army Reserve, *Psychological Operations Support to Strategic Communications in Afghanistan*, Carlisle Barracks, Pa.: U.S. Army War College, strategy research project, March 24, 2009, p. 2. Although it is common to use the terms *adversary* and *enemy* interchangeably, note that, according to Amir Sadat, professor at the National Defense Intelligence College, in his critique of the draft of this monograph, Joint Publication (JP) 3-0 (U.S. Joint Chiefs of Staff, *Doctrine for Joint Operations*, Washington, D.C., Joint Publication 3-0, September 10, 2001) and Field Manual (FM) 3-0 (U.S. Department of the Army, *Operations*, Washington, D.C.: Headquarters, Department of the Army, Field Manual 3-0, February 27, 2008) make the following distinctions: An adversary is a party acknowledged as potentially hostile to a friendly party and against which the use of force may be envisaged (JP 3-0), whereas an enemy is a party identified as hostile against which the use of force is authorized (FM 3-0).

[2]   U.S. Department of Defense, *Information Operations Roadmap*, Washington, D.C., October 30, 2003, p. 6.

We need to understand the people and see things through their eyes. It is their fears, frustrations, and expectations that we must address. We will not win simply by killing insurgents. We will help the Afghan people by securing them, by protecting them from intimidation, violence, and abuse, and by operating in a way that respects their culture and religion. This means that we must change the way we think, act, and operate.[3]

His *ISAF Commander's Counterinsurgency Guidance* emphasized a population-centric strategy that defined popular support for the government, and for the war effort, as a precondition for victory.[4]

During a recent speech, Chairman of the Joint Chiefs of Staff ADM Michael G. Mullen referred to the campaign in Marja as a model of the new counterinsurgency (COIN) strategy, emphasizing obtaining civilian support and restricting the use of force:

We did not swoop in under the cover of darkness. We told the people of Marja and the enemy himself when we were coming and where we would be going. We did not prep the battlefield with carpet-bombing or missile strikes. We simply walked in on time. Because frankly the battlefield isn't necessarily a field anymore. It's in the minds of the people. It's what they believe to be true that matters. And when they believe that they are safer with Afghan and coalition troops in their midst and local governance at their service, they will resist the intimidation of the Taliban and refuse to permit their land from ever again becoming a safe haven for terror.[5]

In another speech, Admiral Mullen expressed concern about the U.S. ability to convince Afghan audiences that they are better off with Afghan and coalition forces:

---

[3] "Is General McChrystal a Hippie?" *Economist*, August 27, 2009.

[4] See GEN Stanley A. McChrystal, *ISAF Commander's Counterinsurgency Guidance*, Kabul: Headquarters, International Security Assistance Force, August 2009.

[5] Charles Lemos, "The Mullen Doctrine," *My Direct Democracy*, March 14, 2010.

It is time for us to take a harder look at "strategic communication."
. . . If we've learned nothing else these past 8 years, it should be
that *the lines between strategic, operational and tactical are blurred
beyond distinction.* This is particularly true in the world of communication, where videos and images plastered on the Web . . .
can and often do drive national security decision-making. . . . We
need to get back to basics. . . . The problem isn't that we are bad at
communicating or being undone by men in caves. Most of them
aren't even in caves. The Taliban and al Qaeda live largely among
the people. They intimidate and control and communicate from
within. . . . No, our biggest problem [is] credibility. Our messages
lack credibility because we haven't invested enough in building
trust and relationships, and we haven't always delivered on our
promises. [emphasis added][6]

Although Admiral Mullen spoke of strategic communication, he
noted that "the lines between strategic, operational and tactical are
blurred beyond distinction," when speaking of the credibility of U.S.
messaging to Afghan villagers, which could just as well be defined as
PSYOP.

Admiral Mullen's comments paralleled those of another senior
official in Afghanistan, U.S. special representative Richard Holbrooke,
who combined IO, PSYOP, and strategic communication, saying that
"the information issue—sometimes called psychological operations or
strategic communication" has become a "major gap to be filled" before
U.S.-led forces can regain the upper hand.[7] In its new strategy for the
Afghan war, the White House has called for an overhaul of strategic
communication in Afghanistan "to improve the image of the United
States and its allies" and "to counter the propaganda that is key to
the enemy's terror campaign." According to the Pentagon's *National*

---

[6]   ADM Michael G. Mullen, chair, U.S. Joint Chiefs of Staff, "Strategic Communication:
Getting Back to Basics," *Joint Forces Quarterly*, Vol. 55, 4th Quarter, August 28, 2009,
pp. 2–4.

[7]   Richard Holbrooke, press briefing on the new strategy for Afghanistan and Pakistan,
Washington, D.C., March 27, 2009. See also Greg Bruno, *Winning the Information War in
Afghanistan and Pakistan*, New York: Council on Foreign Relations, May 11, 2009.

Defense Strategy, "a coordinated effort must be made to improve the joint planning and implementation of strategic communications."[8]

Overlap of Strategic Communication, Information Operations, and Psychological Operations

The cited quotations mixing IO, PSYOP, and strategic communication suggest a lack of consensus on definitions and functions of these activities. Definitions vary for each of these activities, but these definitions are so broad in scope, and so overlapping in functions, that distinctions in practical terms become blurred. *Strategic communication*, for example, is defined as

> [f]ocused [U.S. government] efforts to understand and engage key audiences to create, strengthen or preserve conditions favorable for the advancement of [U.S. government] interests, policies, and objectives through the use of coordinated programs, plans, themes, messages, and products synchronized with the actions of all instruments of national power.[9]

RAND social scientist Christopher Paul, who specializes in IO research, characterizes such a definition as "vague and imprecise. It is not always clear what is, and what is not, part of strategic communication."[10]

---

[8] U.S. Department of Defense, *National Defense Strategy*, Washington, D.C., June 2008, p. 19. See also Bruno, 2009.

[9] U.S. Joint Chiefs of Staff, *Department of Defense Dictionary of Military and Associated Terms*, Washington, D.C., Joint Publication 1-02, April 12, 2001, as amended through September 30, 2010; another definition is "the coordination of Statecraft, Public Affairs, Public Diplomacy [Military] Information Operations and other activities, reinforced by political, economic and military actions, in a synchronized and coordinated manner" (Richard J. Josten, "Strategic Communication: Key Enabler for Elements of National Power," *IO Sphere*, Summer 2006, pp. 16–20, p. 17).

[10] Christopher Paul, "'Strategic Communication' Is Vague: Say What You Mean," *Joint Force Quarterly*, Vol. 56, 1st Quarter, 2010, pp. 10–13, p. 10; also see Christopher Paul, *Whither Strategic Communication? A Survey of Current Proposals and Recommendations*, Santa Monica, Calif.: RAND Corporation, OP-250-RC, 2009.

The official definition of IO in the *Department of Defense Dictionary of Military and Associated Terms* is

> the integrated employment of the core capabilities of electronic warfare (EW), computer network operations (CNO), psychological operations (PSYOP), military deception (MILDEC) and operations security (OPSEC), in concert with specified supporting and related capabilities, to influence, disrupt, corrupt or usurp adversarial human and automated decision-making while protecting our own.[11]

Table 1.1 provides a summary of each of the five basic functions of IO.

During the research for this study, various contrasting explanations were given in interviews concerning what the functions of an IO officer should be, ranging from solely a coordinator of capabilities to the actual implementer of those capabilities. There was general agreement that the IO officer should help translate a commander's objectives into themes and make sure other IO capabilities support the PSYOP process.[12] However, some PSYOP officers argued that IO are superfluous. According to their argument, the PSYOP planning cycle should translate "commanders' objectives into themes," and a separate IO officer is not needed to perform that function.

This doctrinal and bureaucratic conflict within the IO and PSYOP fields was documented in an earlier RAND study:

> RAND conducted interviews with both PSYOP and other IO personnel. . . . We were struck by the startling similarity in the

---

[11] Christopher Paul, *Information Operations: Doctrine and Practice—A Reference Handbook*, Westport, Conn.: Praeger Security International, 2008, p. 2. Paul says that the commander, Combined Arms Center (CAC), has been designated as the Department of Army proponent for building the doctrine, procedures, and techniques to integrate this capability into the larger force and to develop the programs to train personnel in this often-complex strategy. The organization within CAC that is responsible for IO is the U.S. Army Information Operations Proponent (USAIOP).

[12] See Eric V. Larson, Richard E. Darilek, Dalia Dassa Kaye, Forrest E. Morgan, Brian Nichiporuk, Diana Dunham-Scott, Cathryn Quantic Thurston, and Kristin J. Leuschner, *Understanding Commanders' Information Needs for Influence Operations*, Santa Monica, Calif.: RAND Corporation, MG-656-A, 2009.

Table 1.1
U.S. Military Information Operations

IO	Description
PSYOP	Planned operations to convey selected information and indicators to foreign audiences to influence their emotions, motives, objective reasoning, and ultimately the behavior of foreign governments, organizations, groups, and individuals. The purpose of psychological operations is to induce or reinforce foreign attitudes and behavior favorable to the originator's objectives.
MILDEC	Actions executed to deliberately mislead adversary military decision makers as to friendly military capabilities, intentions, and operations, thereby causing the adversary to take specific actions (or inactions) that will contribute to the accomplishment of the friendly mission.
OPSEC	A process of identifying critical information and subsequently analyzing friendly actions attendant to military operations and other activities to: a. identify those actions that can be observed by adversary intelligence systems; b. determine indicators that adversary intelligence systems might obtain that could be interpreted or pieced together to derive critical information in time to be useful to adversaries; and c. select and execute measures that eliminate or reduce to an acceptable level the vulnerabilities of friendly actions to adversary exploitation.
EW	Military action involving the use of electromagnetic and directed energy to control the electromagnetic spectrum or to attack the enemy. Electronic warfare consists of three divisions: electronic attack, electronic protection, and electronic warfare support.
CNO	Comprised of computer network attack, computer network defense, and related computer network exploitation enabling operations.

SOURCE: JP 1-02.

concerns and frustrations expressed by PSYOP and IO representatives. Responses to interview questions were nearly identical. This suggests that, in current operational practice, IO and [their] PSYOP pillar have highly overlapped portfolios. This creates at least three challenges:

There is overlapping authority between IO and PSYOP responsibilities (and potential for consequential animosity).[13]

---

[13] CDR Ed Burns, U.S. Navy, Joint Information Operations Center, interview with Christopher Paul and Todd C. Helmus, Lackland AFB, Tex., February 16, 2006. (Footnote in original.)

There is confusion between respective roles. IO is a staff function and, doctrinally, has strictly a coordinating role. Yet RAND heard anecdotal accounts of IO staffs releasing "IO products" and releasing them without their passing through the rigorous approval process demanded of PSYOP products.

PSYOP's lack of a "seat at the table."[14] With PSYOP subordinate to IO, an IO representative gets direct access to the commander, while PSYOP representatives report to the IO chief. Unless the IO chief is also an expert in PSYOP, this means that relevant shaping expertise in is one step removed from the commander.[15]

In his book *Review of Psychological Operations Lessons Learned from Recent Operational Experience*, Christopher Lamb addressed that latter point:

Many PSYOP officers were irritated by suddenly having to work closely with and through newly designated IO officers. In one case, an IO officer reportedly distributed a product without PSYOP expert input and outside the bounds of the normal PSYOP product approval process, with disastrous effects. . . . The many complaints from PSYOP generally reflected frustration with having to educate untrained IO officers (for example, on the capabilities and limitations of PSYOP).[16]

A major manifestation of this problem is the general substitution of the term *PSYOP* with the term *IO*. According to Paul's research,

Even though the vast majority of PSYOP are based on truthful information . . . because PSYOP aim to influence, PSYOP receives [sic] some of the stigma of propaganda. . . . The growing

---

[14] COL Kenneth A. Turner, U.S. Army, Commanding Officer, 4th Psychological Operations Group, interview with Todd C. Helmus and Christopher Paul, Ft. Bragg, N.C., December 14, 2005. (Footnote in original.)

[15] Helmus, Paul, and Glenn, 2007, pp. 37–38.

[16] Christopher J. Lamb, *Review of Psychological Operations Lessons Learned from Recent Operational Experience*, Washington, D.C.: National Defense University Press, September 2005, pp. 82–83.

pejorative connotation of PSYOP leads to another problem: the conflation of PSYOP with IO. . . . [M]ilitary personnel studiously avoids [sic] mention of PSYOP, instead using the umbrella term. IO has been widely adopted as a euphemism for PSYOP. Consequently, the term IO is now commonly and erroneously used to discuss activities that are, by doctrine, PSYOP.[17]

Douglas Friedly, a senior analyst with the Office of the Under Secretary of Defense for Policy, IO Directorate, contributed the following summation of the problem from the perspective of his Pentagon office (see also Table 1.2):

There has been confusion within DoD, and especially in the field, because the terms SC, IO, and PSYOP (now Military Information Support Operations [MISO]) are frequently used interchangeably. DoD has made a concerted attempt in the last 12+ months to distinguish these terms and convey this distinction, including Congress in various reports. DoD's views of Strategic Communication (SC) is discussed extensively in the Report to the Congressional Defense Committees required by Section 1055 of the FY [fiscal year] 2009 NDAA [National Defense Authorization Act]. Although the term "IO" is often used in a way that implies [that] it is synonymous with individual capabilities such as PSYOP, the term is only appropriately applied when information-related capabilities are coordinated to achieve a military objective. [PSYOP are] the dissemination of information to influence foreign audiences to take action favorable to the U.S. IO describes the integrated employment of a wide range of capabilities to influence adversary decision-making. Although [PSYOP] can be used to inform, persuade, and influence friendly foreign audiences as well as adversaries, PSYOP as part of military activities [are] always integrated into IO.

While there are precise DoD definitions for these terms, all three overlap in the practice of achieving influence. The distinctions have not been of particular concern to commanders, whether

---

[17] Paul, 2008, p. 68.

Table 1.2
Supplemental Chart to Distinguish Strategic Communication, Information Operations, and Psychological Operations

Term	Function	Focus	Comment
Strategic communication	An integrating and coordinating process	Interagency	Totality of U.S. government words and deeds
IO	An integrating and coordinating process	DoD	Consistent with and supports strategic communication
PSYOP (now known as MISO)	A DoD capability	DoD	Supports the commander; always integrated into IO
MIST	PSYOP personnel supporting public diplomacy	U.S. Department of State, public diplomacy	Supports the ambassador in Kabul

SOURCE: Douglas Friedly, senior analyst, Office of the Under Secretary of Defense for Policy, Information Operations Directorate, comments on an earlier draft of this monograph, 2010.

NOTE: MIST = military information support team.

strategic, operational, or tactical, because their interests are typically focused on achieving effects on the ground. Further, the distinction between SC and IO is blurred because in its broadest sense, strategic communication involves the integration of issues of audience and stakeholder perception and response into policy-making, planning, and operations at every level. IO should be consistent with [U.S. government strategic communication] goals and objectives. Where DoD is the lead or a major means to achieve [U.S. government strategic communication] goals among a particular audience, IO efforts may become essentially SC efforts.

Another factor bearing on the confusion in terminology is that influence is not just what is said. It is very much about what is done. While words can be drafted and communicated in very short order, the deeds of individuals, organizations, and even the nation tend to have the strongest and most enduring message that is understood by audiences. A key lesson of the past decade is that what we do is often more important than what we say: the presence of an aircraft carrier in the Gulf, the use of female screen-ers on raids in traditional Muslim communities, or an airstrike on a suspected enemy location all send "messages." The messages received will depend on the audiences, and will not always be the messages we intended to send.[18]

In the field, however, IO are often not seen in the doctrinal terms. In his description of IO in Khost province, Ensign Robert J. Bebber describes a much more hands-on role as an implementer of diverse activities seeking to influence local Afghans, not just a coordinator of core capabilities. Speaking of a research and analytic tool developed to gather information for more-effective creation of themes and messages, Bebber writes,

> the IO Environmental Assessment Tool . . . was designed and used under wartime conditions in an area of the world which can barely be said to be at a Third World level and with more than two-thirds of the population being illiterate.

> The tool was also crafted so that, with a little training, IO planners at the unit level could train squad leaders and troops on its use and purpose, so that other members of the team might also be in a position to collect data for the IO effort. Its use can best be described as *qualitative research*, rather than quantitative. . . . The PRT [provincial reconstruction team] conducted missions almost daily during the time frame it was stationed in Khost (March through November 2008) and the IO officer traveled on the missions most of the time in order to collect data and conduct the assessment. The data [were] updated regularly and reported to the leadership team during the Battle Update Brief (or "BUB"), which was held three times per week. The IO officer was aided by a cultural advisor assigned to the IO unit, a local national working directly for Coalition Forces.[19]

The conflation of IO, PSYOP, and strategic communication is illustrated by Figure 1.1, from an Operation Enduring Freedom (OEF)

---

[18] Friedly, 2010.

[19] Bebber, 2009.

Figure 1.1
Information Operations Means of Dissemination for the 2004 Afghan Presidential Election Campaign

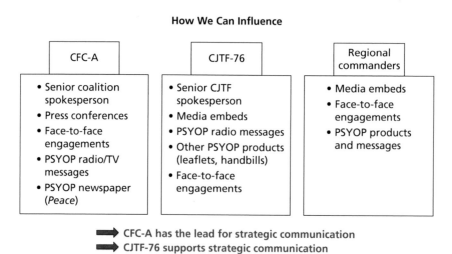

How We Can Influence

CFC-A	CJTF-76	Regional commanders
• Senior coalition spokesperson • Press conferences • Face-to-face engagements • PSYOP radio/TV messages • PSYOP newspaper (*Peace*)	• Senior CJTF spokesperson • Media embeds • PSYOP radio messages • Other PSYOP products (leaflets, handbills) • Face-to-face engagements	• Media embeds • Face-to-face engagements • PSYOP products and messages

➡ CFC-A has the lead for strategic communication
➡ CJTF-76 supports strategic communication

NOTE: CFC-A = Combined Forces Command–Afghanistan.
SOURCE: Operation Enduring Freedom Combined Joint Task Force 76–Afghanistan.
RAND *MG1060-1.1*

Combined Joint Task Force 76 (CJTF-76) PowerPoint briefing on the PSYOP for the 2004 Afghan presidential election campaign plan.

As can be seen in Figure 1.1, most of the IO activities being conducted are within the PSYOP realm, with the exception of press conferences and the activities of the senior coalition spokesperson, which would fall under public affairs (PA). Given this study's PSYOP emphasis in assessing IO, it is useful to review the definition of PSYOP and their wide range of activities. PSYOP are planned operations designed to convey selected information and indicators to foreign audiences to influence their emotions, motives, objective reasoning, and, ultimately, the behavior of foreign governments, organizations, groups, and individuals. PSYOP seek to induce or reinforce foreign attitudes and behavior favorable to the originator's objectives.[20] PSYOP missions

---

[20] U.S. Joint Chiefs of Staff, *Doctrine for Joint Psychological Operations*, Washington, D.C., Joint Publication 3-53, September 5, 2003. There might be conflicting definitions. See also U.S. Marine Corps and U.S. Department of the Army, *Psychological Operations*, Washing-

are "delivered as information for effect, used during peacetime and conflict, to inform and influence,"[21] and fall into three categories:

- "Strategic PSYOP are international information activities conducted by [U.S. government] agencies to influence foreign attitudes, perceptions, and behavior in favor of US goals and objectives during peacetime and in times of conflict."[22]
- "Operational PSYOP are conducted across the range of military operations, including during peacetime, in a defined operational area to promote the effectiveness of the joint force commander's (JFC's) campaigns and strategies."[23]
- "Tactical PSYOP are conducted in the area assigned a tactical commander across the range of military operations to support the tactical mission against opposing forces."[24]

PSYOP functions are to do the following:

- Advise the supported commander through the targeting process regarding targeting restrictions, psychological actions (PSYACT), and psychological enabling actions to be executed by the military force.
- Influence foreign populations by express information through selected conducts to influence attitudes and behavior and to obtain compliance or noninterference with friendly military operations.
- Provide public information to foreign populations to support humanitarian activities, ease suffering, and restore or maintain civil order.

---

ton, D.C.: Headquarters, Department of the Army, and U.S. Marine Corps, Fleet Marine Force Manual 3-53, Field Manual 33-1, February 15, 1993; Headquarters, Department of the Army, *Psychological Operations Techniques and Procedures*, Washington, D.C., Field Manual 33-1-1, May 5, 1994; and JP 1-02, 2001 (2010).

[21] Headquarters, Department of the Army, *Psychological Operations*, Field Manual 3-05.30, Marine Corps Reference Publication 3-40.6, April 2005, p. 1-1.

[22] JP 3-53, 2003, p. ix.

[23] JP 3-53, 2003, pp. ix–x.

[24] JP 3-53, 2003, p. x.

- Serve as the supported commander's voice to foreign populations by conveying the JFC's intent.
- Counter adversary propaganda, misinformation, disinformation, and opposing information to correctly portray friendly intent and actions, while denying others the ability to polarize public opinion and affect the political will of the United States and its multinational partners within an operational area. (FM 3-05.30)

Some views of PSYOP argue that they should be implemented exclusively at the local level, focused on tactical objectives. However, Joint Publication (JP) 3-53 speaks of influencing foreign nations and reaching audiences at a much higher level than villages.[25] It has also been defined as part of strategic communication, capable of achieving strategic deterrence. Limiting PSYOP to tactical support to military operations has become part of the process of overlap of functions with IO and strategic communication.

A separate key point, related to implementation, is that PSYOP should enjoy wide leeway in the manner in which selected information is conveyed and in the specific actions taken to influence the emotions, reasoning, and behavior of target audiences. This can be accomplished through multimedia propaganda, by medical civic action programs (MEDCAPs) and other types of civil affairs (CA) projects, and by face-to-face communication with local leaders. In this respect, it is a mistake to compartment the psychological effect on the civilian population of U.S. military (USMIL) operations exclusively to officially designated PSYOP activities. It can be argued that everything a military force does in a conflictive zone has a psychological impact, favorable or negative, whether intended or not. This is a widely recognized phenomenon described extensively in various studies, such as the following:

> The behavior of every soldier, sailor, airman and marine in a theater of operations shapes the indigenous population. . . . Because of the globalization of media, how a single soldier handles a tactical situation in an out-of-the-way location still has the poten-

---

[25] JP 3-53, 2003, p. ix.

tial to make global headlines and have strategic impact. . . . [I]ndigenous individuals with whom troops interact form favorable or unfavorable impressions . . . and spread those impressions by word of mouth throughout surprisingly large networks.[26]

For example, as current ISAF leadership has admonished repeatedly, the manner in which military convoys drive on the roads has an impact on how those soldiers are viewed, as do hiring practices for locals and myriad other activities. Seen in this context, the everyday activities of troops among the population can have more impact than propaganda disseminated by leaflets[27] or other media. Various statements by commanding officers indicate a clear understanding of this phenomenon, but it is hard to quantify the degree to which this awareness has taken hold among troops.[28]

Adding to the growing body of literature on this issue, MAJ Walter E. Richter comments in his 2009 IO article,

> Kinetic operations involve application of force to achieve a direct effect, such as artillery, infantry, aviation, and armored offensive and defensive operations. Non-kinetic operations are those operations that seek to influence a target audience through electronic or print media, computer network operations, electronic warfare, or the targeted administration of humanitarian assistance. It is important to note that many operations do not fall neatly into one category or the other. For example, a security patrol may have the power to apply force (a kinetic operation), but over time, if its consistently professional conduct earns it the respect of local populace, its presence can become a non-kinetic effect. . . . [T]he difference between kinetic and non-kinetic operations becomes ambiguous. The benefit of this ambiguity is that it allows com-

---

[26] Helmus, Paul, and Glenn, 2007, p. 31.

[27] In this monograph, we use *leaflet* to refer to air-dropped print media and *handbill* to refer to products that are hand delivered.

[28] See McChrystal, 2009.

manders the option of focusing IO on both kinetic and non-kinetic operations. . . .[29]

The concept of every infantryman being a PSYOP operator by virtue of his daily interaction with the locals is illustrated in a compelling manner through various graphics in the 1st Battalion (Bn) 5th Marine briefing, "COIN in Helmand Province: After the Clear—Thoughts and Tips on Non Kinetic Actions,"[30] and in Figure 1.2.

How This Monograph Is Organized

The remainder of this monograph is divided into the following sections:

- a methodology for assessing PSYOP impact (Chapter Two)
- main themes and messages and their effectiveness (Chapter Three)
- a review of means of PSYOP dissemination (Chapter Four)
- an assessment of effectiveness in countering Taliban propaganda (Chapter Five)
- an evaluation of organizational problems of IO and PSYOP (Chapter Six)
- a look at new initiatives being implemented to improve PSYOP (Chapter Seven)
- recommendations for additional changes to improve PSYOP (Chapter Eight).

It also has three appendixes: briefing slides on a campaign plan against improvised explosive devices (IEDs) (Appendix A), briefing slides on a campaign plan to support presidential elections (Appendix B), and a DoD memorandum on the distinctions between strategic communication, IO, and PSYOP (Appendix C).

---

[29] MAJ Walter E. Richter, U.S. Army, "The Future of Information Operations," *Military Review*, January–February 2009, pp. 103–113, p. 104.

[30] 1st Battalion, 5th Marines, "COIN in Helmand Province: After the Clear—Thoughts and Tips on Non Kinetic Actions," undated briefing.

Figure 1.2
Infantry as Psychological Operations Operators

Be nice until it's time to
not be nice. These kids
will be fighting age soon.
Did you help them
choose sides?

Listen to local elders.
And take your gear off
when you're with them.

Put out security, then take
your helmet off when you
talk to people.

SOURCE: 1st Battalion, 5th Marines, undated, slides 70, 81, and 83.
RAND *MG1060-1.2*

CHAPTER TWO

Methodology for Assessing the Effectiveness of U.S. Military Psychological Operations

This monograph tracks the implementation of PSYOP in Afghanistan from late 2001 to 2010. It is an inexact undertaking. Tracking the evolution of specific campaigns in Afghanistan is difficult because there is no central repository of data, neither in the United States nor in Afghanistan, concerning themes and messages disseminated or specific operations and their impact on target audiences. Moreover, IO and PSYOP in Afghanistan have been characterized by a high degree of variation between the different components operating in theater, including special-forces teams in the field, regional commands (RCs), task forces, and the ISAF headquarters in Kabul. What might be an accurate observation for RC East might not apply to RC South. No one has compiled a comprehensive record of all these decentralized PSYOP campaigns. In an effort to protect local collaborators from reprisals and minimize the USMIL public "footprint," PSYOP activities increasingly are classified, which further impedes accurate comparisons of past and present practices.

To assess PSYOP effectiveness, three basic considerations were taken into account, according to the following criteria:

- credibility: This has two major facets: (1) how believable or reasonable the message content is to the target audience and (2) how credible the messenger or means of dissemination is.
- appropriate cultural, social, political, or religious context: This means avoiding the common mistake of *mirror imaging*—that is, presenting propaganda within a U.S. frame of reference, not that

of the target audience. It also includes conforming to Islam as practiced in Afghanistan.

- overall effectiveness: Operations must show evidence that audience perceptions or behavior were influenced as intended.

In conducting these assessments, the study looked at PSYOP themes, products, and actions, which this monograph defines as follows:

- theme: a subject, topic, or line of persuasion used to achieve a psychological objective[1]
- PSYACT: activities conducted for their psychological impact[2]
- PSYOP product: any visual, audio, or audiovisual item generated and disseminated in support of a PSYOP program.[3]

There is abundant literature on the need for an emphasis on cultural appropriateness in PSYOP. It is a basic element of PSYOP doctrine. Moreover, there has been a general awareness of this among U.S. officers serving in Afghanistan throughout the past decade. In an interview about his experiences as the commanding officer of the 3d Battalion, 7th Field Artillery, 25th Infantry Division, Combined Joint Task Force 76 (CJTF-76), based in Kandahar from April 2004 to April 2005, COL Clarence Neason emphasized the importance of tribal culture and politics:

> The local elders of the local shuras were, in fact, the governing piece as you went out to a lot of the remote sites—for that matter, even within some of the towns within Kandahar. As you went and met with them, they spoke for the people, and deference was given to them as we reached out and touched them to find out what their needs were and find out what their [positions were] with regard to the national government. . . . Everything was done

---

[1]   FM 33-1-1, 1994, p. 8-1.

[2]   See FM 3-05.30, 2005, p. 1-2.

[3]   See FM 3-05.30, Appendix A, "Categories of Products by Source," and the product-development information (pp. 1-5, 3-5–3-9, and 6-3–6-9).

along tribal lines. I mean, the tribe was everything in Kandahar. As I dealt with the people there, I had to remain very conscious of, "Am I dealing with the Barakzai or the Popalzai?" You know: Making sure that I am not inadvertently empowering one tribe over the other, because then that would cause friction between them and then you would have problems. [4]

However, MAJ Joseph L. Cox, in his assessment of the OEF campaign during that same period, noted that the cultural training that was given usually consisted of "cultural do's and don'ts" that did little to advance the commander's knowledge of the communication environment in which he operated. Major Cox argued that commanders needed more-detailed knowledge on local religion, family structures, political structures, tribal issues, demographics, cultural norms, mores, and culturally based personal information processing methods to understand what effect his operations would have in an area. This is what General McChrystal has referred to as rich contextual understanding (RCU) and has also been called *human terrain mapping* or *cultural intelligence*.[5] The issue is not whether the military wants to be culturally aware but how good it is at being so. Todd C. Helmus, a behavioral scientist at RAND, and his colleagues make the following relevant observation about a common pitfall: "Cultural assumptions pose a significant threat to shaping operations. Coalition forces have learned the hard way that cultural assumptions are repeatedly proven wrong."[6]

Several examples can be provided of PSYOP initiatives that failed because of lack of understanding of cultural norms and sensibilities. To highlight the un-Islamic ideology and behavior of the terrorists,

---

[4] Christopher N. Koontz, *Enduring Voices: Oral Histories of the U.S. Army Experience in Afghanistan, 2003–2005*, Washington, D.C.: Center of Military History, U.S. Army, 2008, pp. 362–363.

[5] MAJ Joseph L. Cox, U.S. Army, *Information Operations in Operations Enduring Freedom and Iraqi Freedom: What Went Wrong?* Fort Leavenworth, Kan.: School of Advanced Military Studies, U.S. Army Command and Staff College, 2006; interviews with IO personnel; author's personal observations in Afghanistan.

[6] Helmus, Paul, and Glenn, 2007, p. 31.

various leaflets have been designed with Koranic verses printed on them, admonishing the faithful to avoid violence and maintain peaceful relations with everyone. Although the messages themselves were perfectly acceptable, it was questionable in the eyes of the target audience whether unbelievers should be quoting the Koran. Worst of all, these Koranic verses being printed on a leaflet to be dropped from an airplane or a helicopter was not acceptable. It was considered blasphemous to drop pieces of paper with Koranic verses on the ground, because the holy verses of revelation were sullied with dirt. Likewise, U.S. PSYOP personnel dropped colorful soccer balls from low-flying Blackhawk helicopters depicting the flags of coalition nations. However, the Saudi flag has the *shuhada* (declaration of Islamic faith) written on it. Some Afghans and Arabs felt that kicking the holy statement was blasphemy, and the military reportedly apologized to the Saudis and the Afghans for the gaffe.[7]

Mindful of these past mistakes, the present study employs three categories of judgment to rate the effectiveness of PSYOP campaigns: effective, mixed results, or ineffective. These judgments were not made by means of quantitative or mechanistic formula. Given the lack of systematic posttesting or use of surveys or focus groups to more-precisely test target-audience reactions to specific PSYOP products, the study relied on more-subjective judgments based on the following sources: USMIL reporting from the field, including a series of PowerPoint briefings on specific campaigns; press reporting; public-opinion polls; academic studies; interviews conducted with U.S. military and civilian personnel who have conducted IO and PSYOP activities in Afghanistan; interviews with Pashtun tribal leaders and former Taliban members conducted in Kabul during April and May 2009; and interviews with USMIL and NATO officers in Afghanistan during a January 2010 visit.[8]

---

[7] SGM (ret.) Herbert A. Friedman, "PSYOP Dissemination," *psywarrior.com*, undated web page (b).

[8] For the interviews, the Pashtun tribal leaders and former Taliban did not want to be identified by name so as not to be pegged as collaborators with a U.S. effort to influence Afghans. Some of the USMIL personnel interviewed are named in the acknowledgments of this monograph; others preferred anonymity.

Moreover, I have been present at PSYACTs and distributions of PSYOP products and was thus able to directly observe target-audience reactions to the messages. In addition, I interviewed members of target audiences and received frank feedback on how they viewed the credibility and impact of the texts and graphics. I also conducted interviews with Afghans who helped translate and distribute propaganda.

Regarding the use of public-opinion polls, the issue of reliability comes to the fore. Most polls seek to give a picture of the national population in general—that is, interviews are conducted in urban and rural areas, including all the major ethnic groups in Afghanistan. The problem with that approach is that rural Pashtuns in RC East and RC South are the most-relevant population sector for the evaluation of IO and PSYOP effectiveness. It is among them that the Taliban will either win or lose the insurgency. However, national-level polls reflect the opinions of the totality of the population, reflecting groups and regions not germane to the main COIN objective. Increasingly, polling is being done at the district and village levels, some of it sponsored by DoD. This is a big step in the right direction. Nonetheless, these polls generally do not cover reactions to specific messages and themes. Thus, poll results often cited as evidence for effectiveness or ineffectiveness of USMIL messaging need to be viewed with caution.

In the most-recent ABC/BBC/ARD poll conducted in December 2009 and January 2010,[9] there is a reversal of negative trends that had been observed since 2005. This good news was greeted with relief by USMIL and civilian officials in Kabul. This study tentatively accepts the validity of that survey but notes that these findings were based on a sample of only 1,500 respondents from across the country. The small size of that sample and its geographic dispersion call into question whether it should be used at all in evaluating perceptions of rural Pashtun communities. Taking into consideration all these caveats, however, this monograph does utilize these poll results in its assess-

---

[9] See Afghan Center for Socio-Economic and Opinion Research, poll for ABC News, BBC, and ARD, December 11–23, 2009; "Afghanistan: National Opinion Poll for BBC, ABC News and ARD," press release, BBC Press Office, February 9, 2009; and Jill McGivering, "Afghan People 'Losing Confidence,'" BBC News, February 9, 2009.

ments. It also takes into account the more-recent poll results published by the International Council on Security and Development (ICOS)[10] based on interviews with 552 Afghan men in Helmand and Kandahar provinces in June 2010. Depending on the issue, some of their results contradict national-level polling, while others are in agreement. Overall, the trends reported by ICOS are more negative than the earlier ABC/BBC/ARD poll.

It should be emphasized that some IO and PSYOP officers in the field are attempting to conduct their own local public-opinion surveys, which is the best path to follow for ultimately influencing these audiences. This monograph also refers to those very localized findings. On this endeavor, Ensign Bebber, referring to himself as an IO officer, writes the following cautionary notes:

> During the more than 200 interviews, several practices were adopted to elicit more "honest" responses. That being said, it is important to acknowledge up front that interviews are being conducted by an individual in an American uniform, wearing body armor and carrying weapons and with other American and Afghan military and police in the area. Despite the presence of a cultural advisor who was interpreting for the IO officer, some results may have been skewed, but how much or often is unknown.

> There is a Pashtu saying that "A single 'no' is worth a thousand 'yeses.'" This means that whenever questioned by someone, a "yes" response will tend to elicit follow up questions while a "no" response might end the questioning. After decades of brutal Soviet occupation, civil war and the repressive rule of the Taliban, most Afghans are understandably wary when approached and asked if they would mind "just answering a few questions." We must also acknowledge this limitation.[11]

---

[10] International Council on Security and Development, *Afghanistan: The Relationship Gap*, Brussels, July 2010.

[11] Bebber, 2009.

Taking into account the difficulties in gathering reliable data needed to accurately assess the effectiveness of USMIL IO and PSYOP in Afghanistan, this monograph presents a framework of analysis that relies heavily on greater cultural understanding of the target audiences. Figure 2.1 illustrates the framework of the analysis conducted for this study.

The analytic process consisted of first determining major themes based on a review of the sources. Having determined the major themes, examples of specific propaganda products were selected for individual assessment. On this point, some scholars object to the use of the term *propaganda* because it has negative connotations among general audiences, and even within military and policy circles. However, *propaganda* of itself is a neutral term; as defined by DoD, it is "any form of communication in support of national objectives designed to influence the opinions, emotions, attitudes, or behavior of any group in order to benefit the sponsor, either directly or indirectly.[12] There is nothing in the definition to indicate that it is inherently deceptive. On the contrary, propaganda can be completely truthful, depending on the strategy being pursued. Inventing other terms and euphemisms for perfectly good PSYOP terminology that have a long history of thought and practice behind them clouds the issue. The overriding concern with politically careful terminology can be taken as a manifestation of the weak U.S. government stance in the vital arena of propaganda

Figure 2.1
Framework of Analysis

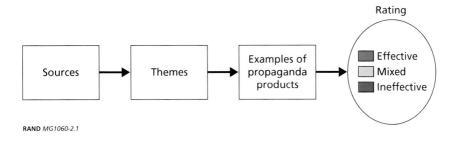

RAND *MG1060-2.1*

---

[12] FM 33-1, 1993.

and counterpropaganda. There was a time when PSYOP were known also as *psychological war.* That is what the Taliban are waging against the United States. That is how they conceive of it, as evidenced by the name of their new English-language website, "In Fight." The United States needs to respond in the same frame of mind, especially when some of its own officers say privately that the information war is being lost.

The products on which this study focused were primarily leaflets, handbills, and posters simply because those products contain written text and graphics and are thus easier to reproduce for the reader. This is not to be taken to mean that these print media are considered to be the main forms of PSYOP dissemination. On the contrary, this monograph has an entire chapter that examines other means of dissemination. Most research on the subject, and the various PSYOP manuals cited in this monograph, consider face-to-face communication—a PSYACT rather than a product—to be the most-effective means of disseminating themes and messages.[13] This is also the main theme of various books and articles by IO and PSYOP practitioners who have written about their experiences, including the detailed account in *Villages of the Moon: Psychological Operations in Southern Afghanistan.*[14]

Once the main symbols, themes, and messages used in the PSYOP products were isolated, the research turned to the sources of information utilized—that is, the interviews, polls, press reporting, and so forth—to find data that would allow making a judgment on the effectiveness of the products. Ideally, the data would be keyed to the product itself, as with the results of a poll or a focus group asked to rate the product. However, that type of data usually did not exist. So it was necessary to extrapolate.

The following example illustrates the process: During the author's interviews in Afghanistan in 2009 with Pashtun tribal leaders and former Taliban, there was a consensus that USMIL night raids, shooting of terrorism suspects in their homes, and killing of civilians during

---

[13] See FM 33-1, 1993.

[14] M. E. Roberts, *Villages of the Moon: Psychological Operations in Southern Afghanistan,* Baltimore, Md.: Publish America, 2005.

air strikes had provoked intense antipathy toward NATO and the U.S. military among Pashtuns in general. One of the interviewees' brother-in-law had been shot to death by U.S. forces inside his house at night, and his sister paralyzed permanently as a result of the gunshots she received. The interviewee kept repeating that "you [the Americans] had entered their bedroom" as a particularly outrageous violation of Pashtun norms in order to commit what he considered to be a criminal act. His anger and desire for revenge are probably shared by his family and clan. We can only speculate how many other Pashtuns who heard his story and sympathize with him (whether or not it is true).

On the other hand, at the November 2010 conference on Afghanistan held in Dubai and organized by the Afghan nongovernmental organization (NGO) the Killid Group, one of the Afghan participants noted that, in the localities where he operates, the Pashtun villagers were very pleased with a recent USMIL night raid because it eliminated a Taliban commander who had recently beheaded two young men from their community and was widely feared. This underscores the dangers of making generalizations based on fragments of data available to evaluate something as complex as perceptions and opinions of a Pashtun target audience. A particular village might be seething with anti–United States sentiment while its neighbors on the other side of the mountain might be seething with anti-Taliban resentment and might look very favorably upon the USMIL presence in their area.

To corroborate the allegations of those who condemned the night raids, this study looked at the polling data, and those data were consistent with the negative interviews.[15] Both interviews and the polls suggest that PSYOP products blaming the Taliban exclusively for violence in the countryside are likely not being well received by some sectors of the target audience because there is such a strong feeling that the U.S. military itself is responsible for much of the violence and for what the locals consider to be terrorist acts—that is, dropping bombs on civilians from the sky and breaking into homes at night to kill people in their beds.

---

[15] The results published by the International Council on Security, 2010, are representative of the negative reporting on Afghan attitudes toward the USMIL presence.

Likewise, PSYOP products extolling the progress that has been brought about by democracy and the new government are undercut by the widely reported, intense disillusionment with the Karzai administration. Nonetheless, in both cases, the effectiveness of the products was rated as mixed because there was also evidence that some sectors of the target audience repudiate the Taliban, do not want USMIL forces to leave soon, and believe that things are getting better in Afghanistan.

The study assumes that it is usually not possible to make cause-and-effect relationships between PSYOP actions and products on the one hand and observed behavior on the other. Other factors, unknown to observers, could account for the particular behavior in question— for example, providing information to the USMIL on IED placements, which might have little to do with the PSYOP leaflet urging that action. This analytic process perforce is subjective. Nonetheless, asking researchers to make judgments on the effectiveness of specific themes and messages, according to the three-part criterion put forth in this monograph, is a valid intellectual exercise that helps understand the complexity of the task at hand. The judgments presented in this monograph should all be considered tentative. However, this process is a concrete means of trying to bring cultural awareness to bear on PSYOP planning in a systematic manner. The judgments made during this study point out questions to ask in pursuing new PSYOP initiatives in order to replicate past successes and avoid past failures and mistakes.

The inherent lack of precision and uniformity in developing PSYOP measures of effectiveness (MOEs) have been noted in previous documents:

> The biggest problem is connecting the shaping action or message with some measurable quantity or quality that is not confounded by other possible causes. For example, many Iraqi soldiers surrendered at the outset of OIF [Operation Iraqi Freedom]. Was this due to PSYOP leaflets dropped instructing them to do so? Was it instead due to the coalition's massive military might? Were there other causes? What was the most likely combination of causes that resulted in the desirable end? In this case, the possible causes are highly conflated, even though the objective being

measured—surrender—is an observable behavior. It would be even more difficult to assess the multiple causes underlying other objectives, such as creating positive public attitudes toward the coalition.[16]

---

[16] Helmus, Paul, and Glenn, 2007, p. 47.

CHAPTER THREE

Main Themes and Messages and Their Effectiveness

Some of the basic PSYOP themes and messages used today date from the beginning of the U.S. intervention in 2001–2002. This affords a good timeline for determining how successful they have been in influencing audiences. Other themes are new, reflecting the fundamental changes in COIN strategy that have taken place and the changed political situation. Overall, however, there is significant continuity in messaging over the past eight years. Most of the actual leaflets and posters reproduced in this chapter are no longer being disseminated, but the messages and themes they illustrate continue to shape the bulk of the propaganda output. That being the case, it is instructive to see the original products in seeking to determine how well these messages and themes have been received.

Some of the early themes have been discontinued or deemphasized. The biggest change has to do with the war-on-terrorism justification for intervening in Afghanistan. With the change in U.S. administrations in 2009, the very term *global war on terrorism* (GWOT) was discontinued, and official documents no longer use that phrase or abbreviation. In the early days, the United States used the terrorist attacks on September 11 to justify its intervention, with poor results in general. Nine years later, that theme has even less resonance among Afghans and is now downplayed. This does not mean that the atrocities of the radicals are ignored. The difference is that the propaganda is currently Afghan-centric, focusing on what the terrorists are doing to the Afghans, as opposed to what was done to the Americans. This is much more effective.

On the other hand, there are new messages—specifically, seeking help from the community in eliminating the threat from IEDs and emphasizing the leading role of the Afghan government and army in making life better for Afghans.

The next several subsections describe and assess the content and delivery of basic IO and PSYOP themes used by the U.S. military over the years, as listed here:

- The war on terror justifies U.S. intervention.
- Coalition forces bring peace and progress.
- Al-Qai'da and the Taliban are enemies of the Afghan people.
- Monetary rewards are offered for the capture of al-Qai'da and Taliban leaders.
- Monetary rewards are offered for turning in weapons.
- Support from local Afghans is needed to eliminate IEDs.
- U.S. forces have technological superiority over the Taliban.
- The Afghan government and the Afghan National Security Forces (ANSF) bring peace and progress.
- Democracy benefits Afghanistan, and all Afghans need to participate in elections.

The monograph describes each theme and provides an assessment of how well it advances U.S. interests by influencing perceptions and behavior of target audiences. Figure 3.1 illustrates the judgments made.

The War on Terror Justifies U.S. Intervention

- credibility: ineffective
- appropriate context: ineffective
- overall rating: ineffective.

From the beginning of the U.S. intervention, USMIL IO and PSYOP used the 9/11 attacks on the United States as the main justification for invading Afghanistan. More recently, in his December 1, 2009, speech at West Point, President Barack Obama declared that

Figure 3.1
Main U.S. Military Themes and Messages in Information Operations

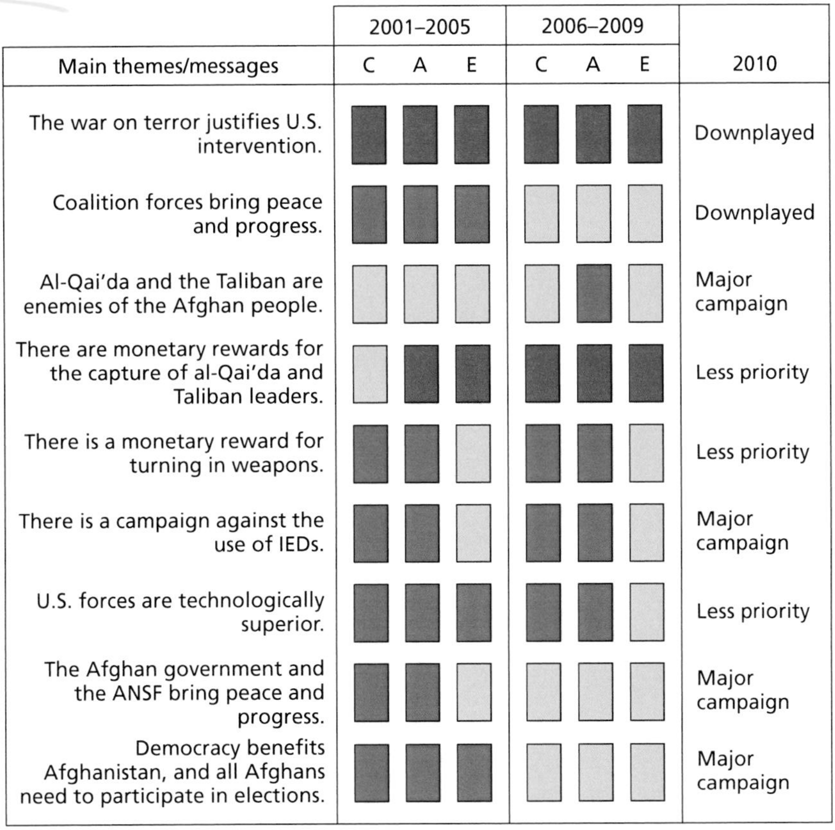

Main themes/messages	2001–2005			2006–2009			2010
	C	A	E	C	A	E	
The war on terror justifies U.S. intervention.	■	■	■	■	■	■	Downplayed
Coalition forces bring peace and progress.	■	■	■	□	□	□	Downplayed
Al-Qai'da and the Taliban are enemies of the Afghan people.	□	□	□	□	■	□	Major campaign
There are monetary rewards for the capture of al-Qai'da and Taliban leaders.	□	■	■	□	■	■	Less priority
There is a monetary reward for turning in weapons.	■	■	□	■	■	□	Less priority
There is a campaign against the use of IEDs.	■	■	□	■	■	□	Major campaign
U.S. forces are technologically superior.	■	■	□	■	■	□	Less priority
The Afghan government and the ANSF bring peace and progress.	■	■	□	□	□	□	Major campaign
Democracy benefits Afghanistan, and all Afghans need to participate in elections.	■	■	■	□	□	□	Major campaign

NOTE: C = whether the theme displays credibility. A = whether the theme illustrates appropriate cultural, social, political, and religious context. E = the theme's overall effectiveness. Green indicates that the theme was effective. Yellow indicates mixed results. Red indicates that the theme was ineffective.

RAND *MG1060-3.1*

the principal reason for remaining in Afghanistan and conducting a "war of necessity" was to keep al-Qai'da from reestablishing a sanctuary that could be used to attack the United States again.[1] The assessment of this study is that, as long as this theme focused on al-Qai'da

---

[1]   Peter Spiegel, Jonathan Weisman, and Yochi J. Dreazen, "Obama Bets Big on Troop Surge," *Wall Street Journal*, December 2, 2009.

and the foreign terrorists who had set up camps in Afghanistan, it had some credibility. It should be noted, however, that most Afghans never saw a terrorist-training camp and had little or no interaction with al-Qai'da. When USMIL IO and PSYOP applied the *terrorist* label to the Taliban, the efforts lost credibility because it appears that most Pashtun target audiences do not consider the Taliban to be international terrorists and do not accept the premise that the Taliban had anything to do with the attack on New York City on 9/11 (despite their alliance with al-Qai'da). Moreover, as the war on terrorism continued in Afghanistan, long after most of al-Qai'da had fled the country and abandoned its camps, this became less credible as a justification for a foreign occupation.[2]

Today, the viability of the war-on-terror theme is further diminished by the fact that there is more terrorism in Afghanistan than ever before, with a continuing increase in Taliban suicide bombings and use of IEDs that kill and maim innocent civilians, paralleling the increase in U.S. troops and combat operations. The Taliban have a strong propaganda campaign arguing that this situation is the fault of the continuing occupation by infidel foreign troops and that, as soon as the foreigners leave, there will be peace. There is a stark war of ideas here: The United States says that it is in Afghanistan to suppress terrorism, whereas the terrorists say that the United States is the cause of terrorism.[3] Insufficient evidence exists to determine which of the two competing narratives has gained the most adherents among target audiences recently. Lack of U.S. credibility on this issue does not automatically translate to credibility for the Taliban.

Current USMIL campaigns focus on a different theme that does have more credibility: the harm that terrorism causes the Afghan people. Generally, 9/11 and the harm that terrorists inflicted on the United States is no longer mentioned. Whatever misfortune befell foreigners years ago is not of as much concern to Afghan villagers as protecting their own communities from growing violence. The theme of

---

[2]   Interviews with Pashtun tribal leaders and former Taliban, 2009.

[3]   Interviews with Afghan journalists and staffs of Afghan research institutes, 2009.

GWOT has been largely discontinued as justification for the continuing intervention in Afghanistan.[4]

Additional comments follow on some specific products.

Propaganda Products Featuring 9/11

To dramatize this theme, posters (such as that shown in Figure 3.2), leaflets, and videos were produced showing the actual attack on the World Trade Center towers in New York City. (Images of the attack on the Pentagon generally were not used in propaganda products,

Figure 3.2
Poster Featuring the 9/11 Terrorist Attacks on the United States

SOURCE: World Trade Center poster AFC035 as presented by Friedman, undated (a).
NOTE: Friedman suggests that the larger text at the top is a date on the Afghan calendar that corresponds to September 11, 2001. He offers a partial translation of the smaller text: "More than 2,800 people were killed, and 3,000 children lost their parents."
RAND MG1060-3.2

---

[4] Interviews with USMIL officers in Afghanistan, January 2010.

probably because it could have been defined as a military target, and that would have elicited much less sympathy from target audiences. A review of propaganda products by opposing sides going back to World War I, conducted by the author of this monograph, indicates that, as a general principle, PSYOP against enemy forces generally focus on civilian, not military, casualties.[5])

In the months after the invasion, U.S. Army PSYOP units produced a video with Dari- and Pashto-language voiceovers, graphics, and Afghan music to spread the military's message about why the United States was in Afghanistan. Reporting from the village of Tadokhiel, Afghanistan, Associated Press reporter Mike Eckel described a health clinic that Army medics had set up, in which villagers watched a video while they waited for a doctor to see them.[6]

Assessment of Effectiveness. According to Eckel, most of the viewers could not understand the images on the screen of airplanes exploding into tall, glittering buildings; well-dressed people running from billowing clouds; firefighters, rubble, dust, and destruction. Up until that point, most of the villagers had never seen a television, and few had ever seen pictures of New York City. Most knew only vaguely that something had happened on September 11.[7] This indicates a basic failure of visual communication, in that the intended audience did not understand the images being disseminated to make the antiterrorist point.

From a cultural perspective, the basic problem was that, although PSYOP planners had assumed that the videos would clarify for Afghans why U.S. forces came to their country, an appropriate target-audience analysis had not been conducted. The same problem was seen in leaflets that depicted the hijacked airliners hitting the twin towers of New York's World Trade Center. A proper target-audience analysis would have revealed that most rural Pashtun audiences had never seen a skyscraper and could not associate the drawings or photographs of the World Trade Center with buildings full of people. Likewise, most of

---

[5]   For example, see the illustrations in Appendixes A and B.

[6]   Eckel, 2002.

[7]   Eckel, 2002.

the target audience had never seen a jet airliner, either, and did not realize that those planes were also full of innocent civilians. Greater attention to the PSYOP planning cycle, explained in the U.S. Army PSYOP handbook,[8] would help to avoid such mistakes. That manual stresses that good target-audience analysis is a prerequisite for developing appropriate messages and themes. This type of analysis should include information on the target audience's worldview and that audience's experience with news and media images.[9]

It should be noted, however, that some accounts, closely duplicating the target audience, media, and messages as critiqued here, claimed excellent results for this theme. One infantry captain, for instance, wrote,

> One of the most profound tools the CA/[PSYOP] group shared with the [Afghans] was a video that [the group] played for villagers on a laptop or portable digital video camera. The video was a compilation of scenes from the events of September 11, 2001, and the days following, with a Pashtun narrative explaining what happened. This proved to be the one thing the Afghans were interested in the most. None of them knew what had happened, and upon seeing the video, they understood and further supported our presence in Afghanistan. The video helped further their dislike of the Taliban and Al Qaida, and support for U.S. forces in Afghanistan grew.[10]

This contrary view to the general negative assessment of the 9/11 PSYOP products underscores the complexity of the PSYOP mission, seeking to influence foreign minds, often without the benefit of accurate, current information on their perceptions, attitudes, and media exposure. It could be that, in different communities, target audiences respond differently to the same themes and messages disseminated in the same manner.

---

[8]   FM 33-1, 1993.

[9]   See FM 33-1, 1993, and FM 33-1-1, 1994.

[10]   CPT Richard Davis, "CA/PSYOPS [sic] in Afghanistan," *Infantry Online*, April 15, 2003.

Propaganda Products Against Osama bin Laden and an Afghan Safe Haven

When the EC-130 Commando Solo[11] began flying over Afghanistan in the fall of 2001, even before the Taliban fell, the war-on-terror justification for U.S. intervention was a main theme, augmented by "coalition forces are friends and bring peace and progress" and "al-Qai'da and the Taliban are enemies and oppressors of the Afghan people."[12] Samples of messages aimed at the civilian population include the following:

- Osama bin Laden and al-Qai'da provoked the United States into a war with Afghanistan through the terrorist attacks of 9/11.
- Only al-Qai'da and their Taliban allies, not the Afghan people, are enemies of the United States.
- Afghans are called upon to help in the war against terrorism.
- Afghans should not allow their country to be used as a safe haven for terrorists. This is sometimes referred to as the "terrorists in your midst" theme and is a key U.S. plea for assistance.

Assessment of Effectiveness. Although public-opinion polls show that most Afghans repudiate terrorism and do not want their country used as a base by al-Qai'da or any foreign terrorist group, it is questionable whether the "terrorists in your midst" theme is or was ever credible to target audiences. During the Taliban regime, al-Qai'da training camps were few and set off in more-isolated areas. Visits to

---

[11] According to a spokesperson for the 193rd Special Operations Wing, the name "Commando Solo" is a hybrid name that refers to commando operations conducted by a single aircraft (Jim Garamone, "U.S. Commando Solo II Takes Over Afghan Airwaves," American Forces Press Service, October 29, 2001).

[12] Although Commando Solo is not flying currently over Afghanistan, this flying PSYOP platform of the 193rd Special Operations Wing, of Harrisburg, Pennsylvania, can broadcast products on AM and FM radio and can send television images over any frequency. Each of the six planes is a flying radio and television station, capable of preempting a country's programming and replacing it with its own broadcasts. Using hourlong formats like commercial stations use, in 2001–2002, the 4th PSYOP Group (Airborne) broadcast news and information, broken up by blocks of Afghan music. See Garamone, 2001; "Commando Solo Radio Scripts: War on Terrorism in Afghanistan," undated; and Marc V. Schanz, "The New Way of Psyops," *Air Force Magazine*, Vol. 93, No. 11, November 2010.

these training sites by local Afghans were discouraged for security reasons, especially as tensions escalated after 9/11 with the anticipation of U.S. retaliation. Some Taliban members did interact with al-Qai'da, but these reportedly were a small minority. In some localities, Arab fighters and other foreigners did mingle with local Afghans, sometimes marrying local women. However, the evidence available suggests that most tribe members inhabiting the vast Afghan landscape never saw an al-Qai'da operative, nor a foreign terrorism trainee, much less Osama bin Laden himself.

Today, al-Qai'da training camps and other visible manifestations of an al-Qai'da presence exist only across the border in Pakistan. The evidence available suggests that, until U.S. forces arrived and proclaimed it, there was little belief among the main Pashtun target audience that Afghanistan had become a safe haven for international terrorists.

Coalition Forces Bring Peace and Progress

- credibility: effective (2001–2005) to mixed (2006–2010)
- appropriate context: effective (2001–2005) to mixed (2006–2010)
- overall rating: mixed.

The themes of friendship and of providing peace and progress are standard wartime themes and messages that many armies have used in the past in a wide variety of conflicts. The corollary message is that the U.S. military wishes no harm to the Afghan people and does not intend to occupy Afghanistan. As a concrete gesture of this altruistic motive for intervening in Afghanistan in 2001–2002, the U.S. military conducted extensive air-drops of food and other supplies to isolated communities along the Afghan-Pakistan border, accompanied by leaflets and radio broadcasts announcing these food drops. Evidently, these gestures were appreciated and helped foment a positive image of the United States at that time. Between 2006 and 2010, effectiveness of this theme declined largely because of growing Pashtun resentment, documented in polls and interviews, of coalition forces' tactics

that they viewed as offensive, including breaking down doors, searching homes, conducting night raids, and bombing villages, all of which
harmed the credibility of USMIL propaganda proclaiming that the
United States had the best interests of the Afghan people at heart. This
seems to be a case of actions speaking louder than words. According to
Thomas Ruttig's study on this issue,

> While the international engagement, both military and civilian,
> was clearly welcomed amongst most Afghans in the first years
> after 2001, more recently they have added their own "condition-
> ality." Demands that Western forces refrain from using harsh
> and culturally insensitive tactics have become widespread and
> public. Some provincial councils and even groups of lower house
> members have boycotted sessions in protest against airstrikes that
> caused civilian casualties. Parliament has demanded legislation to
> regulate the status of foreign forces and for an end to all opera-
> tions in which no Afghan troops are present.[13]

The Issue of Civilian Casualties

The overarching theme that U.S. and coalition forces bring peace and
stability was undercut for years by repeated complaints by President
Karzai over civilian casualties caused by air strikes. This situation
has produced contradictory messaging between the Government of
the Islamic Republic of Afghanistan (GIRoA) and the U.S. government, adversely affecting IO objectives. For example, in May 2009,
Gen. James L. Jones, Barack Obama's national security advisor,
responded to Karzai's public admonition that the United States would
lose the "moral fight" against the Taliban if it kept killing civilians. In
an ABC interview, Jones declared,

> We're going to take a look at trying to make sure we correct those
> things we can correct, but certainly to tie the hands of our com-
> manders and say we're not going to conduct airstrikes would be

---

[13] Thomas Ruttig, *The Other Side: Dimensions of the Afghan Insurgency—Causes, Actors,
an[d] Approaches to "Talks,"* Afghanistan Analysts Network, July 2009.

imprudent.. . . . We can't fight with one hand tied behind our back.[14]

This public U.S. acceptance of collateral damage as an unavoidable part of military operations in Afghanistan played into the Taliban propaganda machine, another example of Karzai being dismissed by his masters as a puppet, even when the issue was the killing of his fellow nationals. All public-opinion polls indicate unambiguously that civilian casualties caused by air strikes are the single biggest complaint among Afghans against coalition and U.S. forces, a complaint echoed by President Karzai to no avail.[15]

When surveyed Afghans gave their opinions on air strikes, their attitudes were overwhelmingly negative. Specifically, the ABC/BBC/ARD survey posed the question, "Do you think the use of air strikes by the U.S. and NATO/ISAF forces is acceptable because it helps defeat the Taliban and other anti-government fighters, or unacceptable because it endangers too many innocent civilians?" In answering this question, 77 percent of respondents said that air strikes are unacceptable, compared with 16 percent who said that they are acceptable. The ABC/BBC/ARD survey also posed the question, "When civilians are harmed in U.S. and NATO/ISAF air strikes, [whom] do you mainly blame: U.S. and NATO/ISAF forces for mistaken targeting, antigovernment forces for being among civilians, or both sides equally?" A plurality of 41 percent gave the first response, and the rest of the sample split evenly between antigovernment forces (28 percent) and both (27 percent). These data suggest that, when launching a kinetic operation, such as an air strike, it would be especially crucial to integrate IO (if this is not already done)—given Afghans' strong tendency to blame the United States, NATO, and ISAF for harming civilians.[16]

---

[14] Brian Knowlton and Judy Dempsey, "U.S. Adviser Holds Firm on Airstrikes in Afghanistan," *New York Times*, May 10, 2009.

[15] See Adam Mynott, "Afghans More Optimistic for Future, Survey Shows," BBC News, January 11, 2010.

[16] ICOS, 2010, p. 31.

The ABC/BBC/ARD poll taken in December 2009 and augmented in January 2010 shows continuing high figures for negative perceptions of air strikes and blame for them. However, reversing a five-year slide, there is a slight improvement in each category. Since there is no PSYOP campaign attempting to justify air strikes specifically, or make them more palatable to Afghan communities, this can be attributed to the new guidance issued by General McChrystal imposing new restrictions on use of air strikes in areas where civilians reside, resulting in a notable decrease in civilian casualties (see Figures 3.3 and 3.4).

More recently, in November 2010, President Karzai condemned NATO night raids, calling outright for a cessation or major reduction of these types of military operations.[17] In the same month, an ISAF spokesperson described the night raids as beneficial.[18] Thus, contra-

Figure 3.3
Blame for Air Strikes, 2009–2010 Comparison

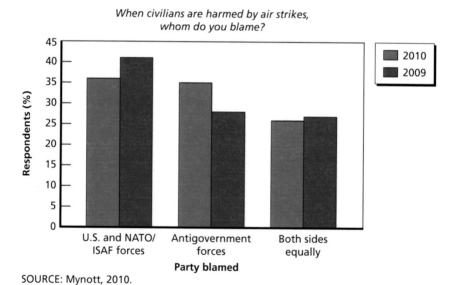

SOURCE: Mynott, 2010.

RAND MG1060-3.3

---

[17] See "Excerpts from Afghan President Hamid Karzai's Interview with *The Washington Post*," *Washington Post*, November 14, 2010.

[18] "ISAF to Continue Night Raids in Afghanistan," *TOLOnews.com*, November 29, 2010.

Figure 3.4
Acceptability of Air Strikes, 2009–2010 Comparison

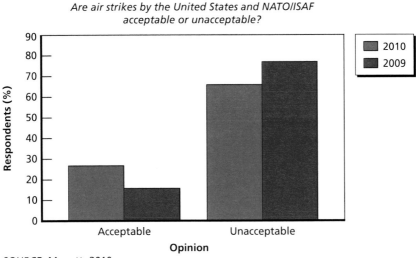

SOURCE: Mynott, 2010.
RAND *MG1060-3.4*

dictory messages continue to be disseminated by GIRoA and ISAF. Instead of a single, reiterated, consistent propaganda theme, USMIL IO and PSYOP had to deal with GIRoA public statements that were more consonant with Taliban propaganda themes than those of the U.S. government. Were it not for evidence that, in some localities, the populace does appreciate the material progress and security brought about by coalition forces, this monograph would have rated the current effectiveness of the overarching theme as ineffective instead of mixed.

With regard to Afghans' attitudes specifically toward U.S. forces, responses across a series of ABC/BBC/ARD polls indicated disillusionment.[19] In particular, Afghans rated the work of the United States in Afghanistan in four surveys conducted in 2005, 2006, 2007, and February 2009. The data from these surveys showed that ratings of the work of the United States plummeted from a high of 68 percent judging this work as excellent or good in 2005 to 32 percent making this

---

[19] "Support for U.S. Efforts Plummets Amid Afghanistan's Ongoing Strife," *Afghanistan: Where Things Stand*, ABC News/BBC/ARD poll, February 9, 2009, p. 1.

judgment in 2009. Conversely, the percentage judging the work of the United States as fair or poor rose from 30 percent in 2005 to 63 percent in 2009. Afghans followed a similar pattern when they judged whether they thought that it was very good, mostly good, mostly bad, or very bad that USMIL forces came into their country to bring down the Taliban government in 2001. These judgments dropped from a high of 87 percent in 2005 claiming that it was very or mostly good to a lower point of 69 percent in 2009. Conversely, the percentage of respondents judging that it was mostly or very bad that U.S. forces entered their country rose from 9 percent in 2005 to 24 percent in 2009.[20]

A similar pattern emerged regarding Afghans' support for or opposition to the presence of USMIL forces in their country today. As with the previous results, a majority of respondents supported the presence of USMIL forces, but that majority decreased from 2006 to early 2009. Specifically, in 2006, 78 percent of respondents either strongly or somewhat supported the presence of USMIL forces, but this number fell to 63 percent in 2009. Conversely, 21 percent of respondents either strongly or somewhat opposed the presence of U.S. forces, but this minority increased to 36 percent in 2009.[21] When it came to confidence levels in the ability of the United States, NATO, or ISAF to provide security and stability in their area, Afghans followed a starker trend. In 2006, most respondents (67 percent) said that they were very or somewhat confident in these forces' ability to provide security, but, by 2009, this percentage dropped to a minority of 42 percent. Conversely, a minority of respondents (31 percent) said in 2006 that they were not so confident or not at all confident in these forces to provide security; by 2009, a majority of 55 percent said that they were not confident in these forces to provide security.[22]

When pollsters asked Afghans in 2005 how favorably or unfavorably they viewed the United States, 83 percent had a very or somewhat favorable view. However, that percentage dropped to a minority

---

[20] "Support for U.S. Efforts Plummets Amid Afghanistan's Ongoing Strife," 2009, p. 22.

[21] "Support for U.S. Efforts Plummets Amid Afghanistan's Ongoing Strife," 2009, p. 8.

[22] "Support for U.S. Efforts Plummets Amid Afghanistan's Ongoing Strife," 2009, p. 26.

of 47 percent by 2009.[23] In parallel, the percentage of respondents who held a somewhat or very unfavorable view of the United States constituted a minority of 15 percent in 2005, but that percentage jumped sharply to a 52-percent majority by 2009.[24]

The poll illustrated in Figure 3.5 shows a steady decline from 2005 to 2009 in how the work of the United States in general was rated in Afghanistan. The slight reversal of this decline in 2010 is probably attributable to the fact that General McChrystal made a systematic effort to address long-standing complaints about coalition forces' tactics and has issued new guidelines aimed at avoiding further antagonism.

Figure 3.5
Afghan Poll Rating U.S. Work, 2005–2010

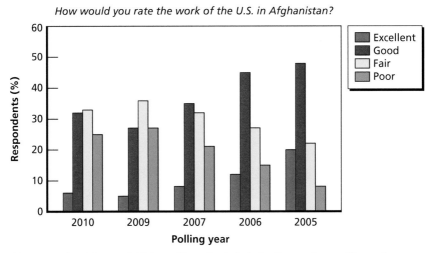

SOURCE: Adam Mynott, "Afghans More Optimistic for Future, Survey Shows,"
BBC News, January 11, 2010.
NOTE: No survey was conducted in 2008.
RAND MG1060-3.5

[23] "Support for U.S. Efforts Plummets Amid Afghanistan's Ongoing Strife," 2009, p. 1.

[24] "Support for U.S. Efforts Plummets Amid Afghanistan's Ongoing Strife," 2009, p. 2.

Handshake Leaflets and Posters

In conjunction with the radio broadcasts of this theme, one of the first leaflets dropped in Afghanistan (Figure 3.6) depicts a U.S. soldier shaking hands with an Afghan citizen, with text written in Pashto and Dari that says, "The partnership of nations is here to help Afghans."[25]

Assessment of Effectiveness. In Afghanistan, it is evident that these leaflets, handbills, and posters did have wide reach and were very effective initially because the Afghans were war-weary, disillusioned with the Taliban regime, and ready for change. They had high hopes that the U.S. intervention would bring peace, progress, and security to Afghanistan, as USMIL propaganda promised. Judging by the lack of widespread resistance to the occupation of Afghanistan (as dramatized by the rapid fall of the Taliban regime with few Afghans rushing to its

Figure 3.6
Leaflet Emphasizing a Coalition/Afghan Partnership

SOURCE: JP 3-53.
RAND *MG1060-3.6*

---

[25] JP 3-53.

defense), the Afghan target audience seemed to have accepted the overall theme that the partnership of nations had come to help Afghans.

American and Afghan Families Juxtaposed

To counteract al-Qai'da and Taliban propaganda that U.S. invaders were infidels who were on a crusade against Muslims, USMIL PSYOP sought to humanize the U.S. image by showing an American family juxtaposed with an Afghan family (Figure 3.7).

Assessment of Effectiveness. From a cultural and historical perspective, the handshake symbol was effective in Afghanistan at that time because it had been used previously as the U.S. Agency for International Development (USAID) symbol in the late 1960s and 1970s, when it built major irrigation-infrastructure projects that most Afghans viewed favorably. To many Afghans, the clasped hands symbolized past U.S. willingness and capacity to build up their country and fueled expectations that the United States would do so again. It is not known whether the designers of this leaflet were aware of this concrete, posi-

Figure 3.7
Leaflet Emphasizing U.S.-Afghan Friendship

SOURCE: "Friendship II?" leaflet AFD030b, as presented by Friedman, undated (a).
NOTE: Friedman translates the text above the clasped hands as "Friendship"
and says that *Army Times* reported this leaflet drop on November 19, 2002.
RAND *MG1060-3.7*

tive connotation in Afghan eyes, or whether they were simply seeking to denote friendship.

Good Times Have Returned to Afghanistan

The leaflet depicted in Figure 3.8 states that the installment of a new government in Afghanistan would bring new liberties and benefits.

Assessment of Effectiveness. From a religious perspective, the crescent moon symbol suggests that this new regime would be Islamic, but the musicians at top left indicate that the Taliban's extremist fundamentalist restrictions on traditional music and other traditional Afghan cultural forms would be eliminated. (The use of the crescent moon is an example of good use of a symbol meaningful to the target audience.) From a social and cultural perspective, the smiling girl also suggests a better deal for women, but this raises the question of who the target audience is. Were Afghan women a significant target audience at that time? Did their views matter? Some observers argue that women are indeed a key audience. Afghan feminists argue that, even though women's traditional domain is confined to the household, as mothers and wives, they do have the power to influence the men in

Figure 3.8
Leaflet Showing Peace and Prosperity as a Result of a New Regime

SOURCE: Reverse of leaflet AFG105, as presented by Friedman, undated (a).
RAND MG1060-3.8

their family. That interpretation of a Pashtun household is unproven, however, as there have been no systematic studies on the subject. Furthermore, there is no evidence available from focus groups or public-opinion surveys as to how well this type of leaflet was received, but it certainly does seem to fit the mood of the time.

Peace and Friendship

Figure 3.9 depicts a leaflet on the peace-and-progress theme combining the clasped-hand and dove symbols against the background of Afghanistan with the Afghan flag color scheme. The white dove is used universally as a symbol of peace and, in Afghanistan, it probably had an additional positive connotation for PSYOP objectives in that raising white doves is a traditional Afghan custom, which, according to various Afghans interviewed for this study, was banned by the Taliban. It should be noted, however, that the colors on the Afghan flag run vertically, not horizontally as shown in the leaflet.

Figure 3.9
Leaflet Emphasizing Peace and Friendship

SOURCE: Front of leaflet AFG105, as presented by Friedman, undated (a).
NOTE: Friedman offers the text translation, "A new government offers new freedoms. The future of Afghanistan depends on your support of the new government."
RAND MG1060-3.9

Assessment of Effectiveness. At present, it appears that the themes of peace and progress lack credibility. Propaganda does not exist in a vacuum; there has to be some connection to reality. Most Afghans consider that Afghanistan is suffering from more violence and bloodshed today than when U.S. troops arrived in 2001. Therefore, target audiences are not likely to accept optimistic propaganda that conflicts with the reality they see around them.

That said, the situation is complex, and Afghans exhibit highly contrasting opinions on basic issues pertaining to the state of their nation. On the one hand, they might be very critical of the government, the Taliban, and foreign military forces but, at the same time, express optimism that things will get better. Moreover, public-opinion surveys usually indicate that most Afghans believe they are better off now than they were under Taliban rule. In the poll illustrated in Figure 3.10, there was a general slide between 2005 and 2009 in terms of the perception that Afghanistan was going in the right direction. This affected campaigns proclaiming that things were getting better. There is a reversal of the downward trend in 2010, but it is premature to predict whether this is temporary or reflects a new trend.

Nonetheless, despite the 2010 spike in positive views, there has been a palpable disenchantment in the past several years with the Afghan government and its handling of the general welfare of the people—a disenchantment that cannot easily be erased. The facts observed on the ground do not conform to the propaganda; therefore, the latter lacks credibility. Widely publicized complaints over corruption damage the regime's legitimacy. It should be noted that, in the DoD assessment for Congress, *Report on Progress Toward Security and Stability in Afghanistan*, there is documented material indicating slow but palpable improvement in various aspects of the Afghan situation.[26] However, this good news is offset by the conclusion that "the popula-

---

[26] U.S. Department of Defense, *Report on Progress Toward Security and Stability in Afghanistan: Report to Congress in Accordance with the 2008 National Defense Authorization Act (Section 1230, Public Law 110-181), as Amended, and United States Plan for Sustaining the Afghanistan National Security Forces: Report to Congress in Accordance with Section 1231 of the National Defense Authorization Act for Fiscal Year 2008 (Public Law 110-181)*, Washington, D.C., April 2010b.

Figure 3.10
Afghan Opinions on the Direction of Their Country, 2004–2010

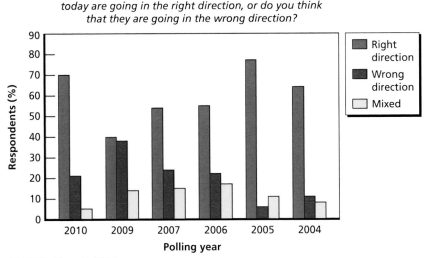

SOURCE: Mynott, 2010.

RAND *MG1060-3.10*

tion sympathizes with or supports the Afghan Government in 24% (29 of 121) of all Key Terrain and Area of Interest districts."[27] Such a low figure for government support suggests that the PSYOP campaign to promote that support is not having its intended effect.

Although not officially part of PSYOP, CA projects and varied types of humanitarian assistance and reconstruction projects definitely enhance the ability to influence local Afghans and convince them to support the Afghan government and the U.S. military and reject the Taliban. Actions speak louder than words, and it is often much more effective to provide a village with concrete benefits that they appreciate than to hand out propaganda leaflets or direct a radio broadcast in their direction.

IO and PSYOP officers debriefed upon returning from the field mention in particular two CA-type projects: (1) the positive effect of

---

[27] DoD, 2010b, p. 7.

MEDCAPs in winning hearts and minds and (2) using the Commanders' Emergency Response Program (CERP) funds to rebuild schools and other facilities desired by local villagers. Projects pursued include providing electricity to villagers, buying desks for schools, providing food aid, outfitting a medical clinic, and even paying for solar-energy panels for the roof of a school (at the request of the locals themselves, who had already experimented successfully with solar panels). Another staple, used effectively all over Afghanistan, is the digging of wells.[28] IO officers pick the villages they want to help, depending on their operational requirements and where they feel they need to make a special effort to gain goodwill and, they hope, cooperation and intelligence for conducting operations.[29] However, the IO officers often do not perform all the administrative and logistical tasks associated with these kinds of projects; often, these tasks are completed by PSYOP or CA personnel. In one location, the standard procedure was for the IO officer to identify a project then turn it over for funding and execution to the local PRT.[30]

CPT Richard Davis speaks in detail of the humanitarian projects with which he was involved, then makes a direct link to success in force protection:

> Through these meetings, the CA/[PSYOP] group was able to aid the local villages in improving their quality of life in many ways as a means of assisting in force protection. The group helped the [Afghans] plan and build an irrigation system. . . . At schools, they would hand out pens, paper, crayons, benches [that] they had purchased, and other school supplies. At local hospitals or clinics, they would [hand out] medications, blankets etc. While the CA/[PSYOP] members were visiting with villages, especially

---

[28] Abundant reporting exists on digging of wells in Afghanistan as a CA priority, including "Joint Group Brings Aid to Villages in Helmand," press release, International Security Assistance Force Afghanistan, December 26, 2010; and "Villagers Seek Medical Help from ISAF Camp," *Sada-e Azadi Radio*, May 23, 2010.

[29] "Commander of Army Civil Affairs and Psychological Operations Command Leaving for New Job," Associated Press, August 20, 2009; interviews with IO personnel.

[30] Interview with IO officer who served in Afghanistan in 2009.

schools, they would attempt to prevent injury to children from unexploded [ordnance] (UXO) or land mines. They would pass out and explain UXO warning posters. . . . [T]he CA/[PSYOP] group ventured out into the surrounding villages and held meetings with the locals. During these meetings, the CA/[PSYOP] group would build rapport with the [Afghans]. By demonstrating genuine concern and giving the villagers humanitarian aid, they quickly developed a feeling of trust with the [Afghans]. Through this relationship, the CA/[PSYOP] group became people with whom the [Afghans] felt comfortable giving information to regarding the locations and plans of Al Qaida and Taliban sympathizers. On numerous occasions, individuals or groups of [Afghans] would approach the CA/[PSYOP] group with information as to where Taliban and Al Qaida members or sympathizers were. In addition, Afghan sources would approach the CA/[PSYOP] group and tell them of expected attacks on the FOB [forward operating base] or U.S. patrols. This only occurred with the CA/[PSYOP] group because the locals felt comfortable approaching them. Never did the locals approach an infantryman to divulge information.[31]

Captain Davis' account is remarkable because everything turned out so well, "by the book." However, other individuals have had much more-clouded experiences in trying to conduct IO and COIN in tribal areas. Interviews with PSYOP advisors returning from Afghanistan indicate that many CA programs are ineffective because units can build schools and clinics but have no money to fund teachers or doctors, provide training, or obtain supplies. The IO effect of this shortcoming is described as profound. The buildings go unused—or worse, become targets for the Taliban to attack and burn. This obviously does not strengthen Afghan confidence in the U.S./coalition commitment to their country.

One officer who recently returned from eastern Afghanistan noted that everyone he knew in the villages appreciated the building of schools and wells but that what they wanted most was security. This

---

[31] Davis, 2003.

goes to the issue of appropriate cultural, social, and political context, as well as credibility. Public-works and humanitarian projects will not achieve their intended effect, he argued, if the Taliban continue to operate with impunity in the area. In his own case, he related the demoralizing impact of the Taliban's assassination of his unit's local translator, in the translator's home village. News of the killing of those who work with Americans—special targets for the Taliban—spreads quickly throughout the countryside and neutralizes whatever goodwill the humanitarian projects might have garnered.[32]

In his article on the future of IO, Major Richter makes a strong argument for a more-standardized, or institutionalized, integration of IO with the CA activities described in this monograph:

> Experiences in Afghanistan further demonstrate the need to integrate public affairs and civil affairs into information operations. . . .[33] PSYOP provided support to the . . . humanitarian de-mining operations. Civil affairs Soldiers also coordinated with non-governmental organizations as part of the State Department's Overseas Humanitarian Disaster and Civic Aid program. . . .[34] These experiences highlight the integral role CA has already played in successful IO as a means to influence the populace. The potential of proper CA integration is not the ability to "win hearts and minds." Rather, it is the ability to establish relationships of mutual respect and trust that foster popular support as all sides recognize the long-term benefits of cooperating with coalition forces.[35]

---

[32] Interviews with IO officers who had returned from Afghanistan.

[33] "Operation Enduring Freedom—Afghanistan," *Global Security*, undated. (Footnote in original.)

[34] BG Pat Maney, Deputy Commander, USACAPOC (A), "Lessons Learned in Afghanistan Slideshow," National Defense Industrial Association—SO/LIC Symposium, February 11, 2003. (Footnote in original.)

[35] Richter, 2009, pp. 103–113. Richter also footnotes LTC Charles Eassa, Deputy Director, U.S. Army Information Operations, as quoted in Michael Schrage, "Use Every Article in the Arsenal: Good Press Is a Legitimate Weapon," *Washington Post*, January 15, 2006.

In his IO and Afghan COIN article, CDR Larry LeGree emphasizes the utility of agricultural development teams as CA support for IO:

> The contributions of the U.S. Department of Agriculture (USDA) and the National Guard's Agriculture Development Teams in Afghanistan cannot be overstated. They have had perhaps the greatest impact per person in Afghanistan. Local farmers who work with USDA representatives to improve the technical aspects of their productivity can improve their yields about 30 percent immediately. The impact of this is huge. The farmer not only has enough food to feed his immediate family—his most pressing need and what he cares about—but also has an excess of food. Now he has the ability to trade and buy and sell goods. The secondary and tertiary effects lead to increased demands for goods in the local markets, sparking further demand for imports and services, and attacking the cycle of poverty. By integrating this type of message into your IO campaign, you become relevant to the right people. You showcase what the insurgents cannot offer.[36]

The early leaflets reproduced for this monograph captured the positive mood of the moment when they were disseminated and helped promote a sense of hope, but the situation has changed fundamentally. In those places where communities have benefited from projects carried out by international aid agencies, or the U.S. military directly, through PRTs and other mechanisms, a favorable disposition to such propaganda could be expected. Unfortunately, too many communities have been left out of assistance programs and have not experienced palpable benefits in the current regime. Thus, a large part of the target audience today might react to the upbeat messages of these types of leaflets with disbelief or cynicism.

This cynicism is evident in the results of Ensign Bebber's Khost survey. Bebber goes on to make some observations of the value of civic action or community assistance projects for IO purposes, which are insightful and relevant to Afghan COIN in general:

---

[36] LeGree, 2010, p. 26.

It appears that previous Coalition units might not have been effective in managing expectations on the part of the local population in terms of development. One common refrain is, "You have been here for six years. How come you have not done more? How come we still live like this?" While there has been much publicity on the fact that the local government and Coalition Forces have spent approximately $50 million on reconstruction and development in Khost last year and this year, it might as well have been $50 *billion* if the vast majority of the local population does not see a tangible benefit in their daily life.

To be sure, there are tangible benefits that are a direct or indirect result of the international presence. During the Taliban regime, there were no cell phones, televisions, barely any paved roads, few schools and even fewer clinics and hardly any personal automobiles. The local population seems to recognize this.

Continued emphasis on reconstruction and development by the local government and Coalition Forces will further legitimize the current Karzai regime and maintain popular support, but it is important to remember that this probably has a *diminishing* utility with the passage of time. Taken in context with the other measures of the environment (i.e., a deteriorating security environment) its effect is further compromised.

One mitigating factor in this is that Afghanistan has just emerged from decades of war and tyranny. The massive social upheaval has helped undo many of the cultural and traditional bonds that were the "glue" of the Afghan people. Coming out from under the shadow of Soviet occupation, civil war and the radical despotic rule of the Taliban, it is not unsurprising that people's expectations are easily inflated as to the capability of highly developed and wealthy nations to radically transform and improve the infrastructure and standard of living. This was true in Iraq (a more developed nation than Afghanistan) whose population was initially frustrated with the pace of reconstruction after the U.S. invasion of 2003. It was also true of Russia after the fall of communism, where expectations failed to meet the reality of the daunting task ahead. Therefore, from an IO perspective, we should not be surprised at some level of disappointment being a constant factor.[37]

Mistrust of foreign presence is suggested in Table 3.1, which shows responses to the June 2010 International Council on Security and Development (ICOS) survey in Helmand and Kandahar provinces. A basic function of PSYOP is to justify military operations to the civilian population, but, in the ICOS survey, the majority viewed recent U.S. and coalition forces' operations in Helmand and Kandahar negatively and opposed the new offensive focusing on Kandahar City and its environs (see Figures 3.11 and 3.12 and Tables 3.2 and 3.3). The ICOS survey results have recently been corroborated by news reports of complaints about demolition of houses and destruction of crops and trees by the ongoing campaign against the Taliban in Kandahar. These complaints are being exploited by Taliban propaganda, as illustrated in a YouTube video mixing interviews with local Kandaharis condemning

Table 3.1
What Are the Foreigners Fighting For?

Response	Helmand (%)	Kandahar (%)	Overall (%)
To occupy Afghanistan	24	10	18
For their own targets (al-Qai'da)	17	12	15
For violence and to destroy Afghanistan	20	6	14
For their own benefit	9	20	14
Don't know	10	18	13
Peace and security	5	21	12
To destroy Islam	12	4	9
No answer	1	6	3
Rebuilding Afghanistan	2	0	1
Other	0	2	1

SOURCE: ICOS, 2010, p. 27.

---

[37] Bebber, 2009. Emphasis in original.

Figure 3.11
Popular View of Military Operations

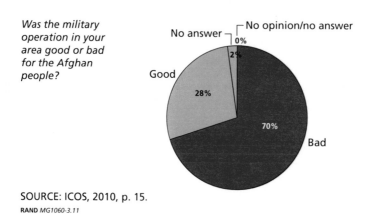

Was the military operation in your area good or bad for the Afghan people?

SOURCE: ICOS, 2010, p. 15.
RAND *MG1060-3.11*

Figure 3.12
Popular View of the Offensive Against the Taliban

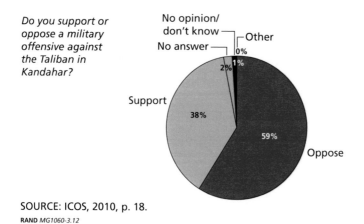

Do you support or oppose a military offensive against the Taliban in Kandahar?

SOURCE: ICOS, 2010, p. 18.
RAND *MG1060-3.12*

the destruction caused by USMIL and coalition forces, with photographs of Afghan men being detained by U.S. soldiers and other scenes of civilian suffering.[38]

---

[38] Larawbar, untitled video, January 12, 2011.

Table 3.2
Popular Views of Military Operations

Was the military operation in your area good or bad for the Afghan people?			
Area	"Good" (%)	"Bad" (%)	No Answer (%)
Garmsir	23	77	0
Marjah	1	99	0
Nawa	20	80	0
Lash City	27	69	4
Helmand	16	83	1
Kandahar	44	54	2
Kandahar City	64	33	3
Khakrez	16	84	0
Panjwayi	24	73	3
Spin Boldak	40	58	2
Total	28	70	2

SOURCE: ICOS, 2010, p. 16.

Al-Qai'da and the Taliban Are Enemies of the Afghan People

- credibility: mixed
- appropriate context: mixed (2001–2005) to effective (2006–2009)
- overall rating: mixed.

PSYOP themes and messages seek to draw a sharp distinction between U.S. forces (seeking to bring peace, stability, and progress to Afghanistan) and al-Qai'da and the Taliban (spreading violence and destruction). Some messages focus on specific acts of destruction that the Taliban perpetrate, such as burning girls' schools and attacking Afghan and foreign aid workers implementing diverse projects that benefit the community. According to interviews with Pashtun tribal

Table 3.3
Popular View of the Offensive Against the Taliban

Do you support or oppose a military offensive against the Taliban in Kandahar?				
Area	Support (%)	Oppose (%)	Other (%)	No Answer (%)	No Opinion/ Do Not Know (%)
Garmsir	46	52	0	1	1
Marjah	20	80	0	0	0
Nawa	35	61	0	2	2
Lash City	42	56	0	2	0
Helmand	35	63	0	1	1
Kandahar	42	53	1	3	1
Kandahar City	68	27	1	2	2
Khakrez	4	91	0	4	0
Panjwayi	6	89	0	5	0
Spin Boldak	48	50	2	0	0
Total	38	59	0	2	1

SOURCE: ICOS, 2010, p. 18.

leaders, as well as returning USMIL personnel from the field, propaganda highlighting Taliban acts of destruction and terrorism seem to be effective. This propaganda plays on resentment of local villagers against the Taliban for making their lives worse. Also, there is fear of the Taliban because of their beheadings of those whom they define as spies or collaborators with the government and the infidels. In every national poll taken, the Taliban are generally seen as the worst problem facing Afghanistan. However, it is important to stress that, depending on the location, district-level polling could yield contrary results. Some localized polling shows strong support for the Taliban (see Figure 3.13).

Consistent poll results, such as these, lead some analysts to conclude that the Taliban is losing the information war. Indeed, some studies suggest that Taliban messaging is not credible, except when

Figure 3.13
Afghan Polls, 2005–2010: Biggest Danger to Afghanistan

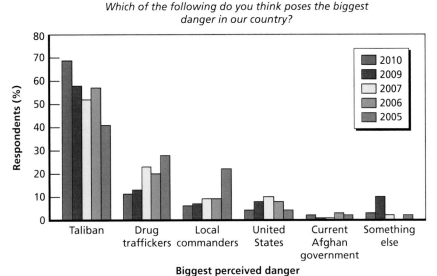

SOURCE: Mynott, 2010.

RAND *MG1060-3.13*

it comes to threats. For example, in the ICOS June 2010 survey in Helmand and Kandahar provinces, 72 percent of the respondents said that they would prefer their children to grow up under an elected government rather than under Taliban rule.[39] Nonetheless, it is hard to believe that a movement as resilient as the Taliban, with a continuing ability to attract new recruits to replace its losses, functions only on the basis of threats. On the contrary, there is strong evidence that the Taliban do manipulate Pashtun cultural and religious traditions to their advantage. For example, UK Strategic Communication Laboratories researchers, based on their interviews conducted in 2010 in Kandahar province, conclude that the Islamic messaging of the Taliban— depicting themselves as true jihadists defending Islam and Afghanistan

---

[39] ICOS, 2010.

from foreign invaders—is highly effective among the Pashtun target audiences.[40]

Furthermore, whatever lack of credibility Taliban propaganda might have on certain points does not automatically translate into USMIL credibility. On the contrary, target audiences seem to be wary of the propaganda coming from both sides. The biggest hurdle to USMIL credibility, as mentioned previously, is those tactics that antagonize the civilian population and that the civilian population sees as abusive. Even in communities that have not experienced air strikes, house searches, and nighttime raids, the perception exists that such assaults can happen at any time. Afghan ambivalence on this issue is exemplified in the following quotation from Mohammed Ishaq Khan, an Achakzai Pashtun tribal leader: "Ten percent of the people are with the Taliban, 10 percent are with the government, and 80 percent are angry at the Taliban, the government and the foreigners."[41] Although the respondents to the ICOS survey overwhelmingly rejected the prospect of the return of a Taliban government, they also expressed a set of negative views about USMIL and NATO forces that are consistent with Taliban propaganda claims:

> For instance 75% of interviewees believe that foreigners disrespect their religion and traditions; 74% believe that working with foreign forces is wrong; and 68% believe that NATO-ISAF does not protect them. [Fifty-five percent] of interviewees believe that the international community is in Afghanistan for its own benefit, to destroy or occupy the country, or to destroy Islam.

> These results are troubling, and demonstrate the mistrust and resentment felt towards the international presence in Afghanistan. Of those interviewed, 70% believe that recent military actions in their area were bad for the Afghan people, whilst 59% opposed further operations in Kandahar. According to interview-

---

[40] Strategic Communication Laboratories presentation at RAND Washington office, November 10, 2010.

[41] Kathy Gannon, "Afghans Blame Both US, Taliban for Insecurity," Associated Press, April 16, 2010.

ees, the Afghan government is also responsible by failing to provide good governance. [Seventy percent] of respondents believe that local officials make money from drug trafficking, and an astonishing 64% state that government administrators in their area were connected to the Taliban insurgency.[42]

In DoD's own periodic assessment of progress in Afghanistan, its comparison of key districts between December 24, 2009, and March 18, 2010, found that districts that either supported or sympathized with the insurgents increased from 33 to 48 percent. This suggests a setback in PSYOP efforts to depict the Taliban as the enemies of the Afghan people, at least in those particular districts, and helps account for the overall mixed assessment of effectiveness for that theme.[43]

In his assessment of IO and COIN, Commander LeGree cautions against a propaganda focus on denigrating the enemy:

> This is not the time to fall prey to the trap of cognitive dissociation—the inability to see perspectives other than one's own. Target audience analysis fails if countering the enemy is the primary preoccupation. The concerns of the average citizen on an average day should be the basis for the IO campaign. . . .
>
> When the IO campaign's radio spots, billboards, and public announcements exclusively focus on reporting improvised explosive device (IED) incidents, offer rewards for information about insurgents, or make clumsy attempts to paint the insurgents as bad guys, the audience is not interested. These things are simply not what the average Afghan cares about. It just gives the insurgents "free press." Tell a man how to grow more wheat on his small plot, give him access to a wider variety of food, or tell him about the bridge that will let him walk to a market and you have the audience's attention. These are the things that matter, the most effective subjects for the IO campaign.[44]

---

[42] ICOS, 2010, p. 2.

[43] DoD, 2010b.

[44] LeGree, 2010, p. 26.

Multimedia Products Disseminating the Message That the Actions of al-Qai'da and Taliban Terrorists Are Un-Islamic

This is an attempt to take the bull by the horns and accuse terrorists who claim to be Islamic holy warriors of not being Islamic—that is, of violating Islamic rules of just war. This is a refrain often voiced by the Afghans themselves as they contemplate the growing carnage in their country being brought about by suicide bombers that often kill civilian bystanders.[45]

Assessment of Effectiveness. Although this is potentially a very powerful theme, and one used by anti-Taliban Afghans themselves, it is a difficult one for the U.S. military to promulgate credibly. Many Muslims believe that the actions of Islamic terrorists are indeed un-Islamic, but, in the endeavor to manipulate minds in the Middle East and in other parts of the world, who says something is as important as what is being said. In this case, U.S. personnel perceived by the Muslim target audience as unbelievers have little credibility for declaring the actions of any Muslim to be un-Islamic. The same message would have much greater effect coming from the mouth of a Muslim key communicator, such as a local mullah. An indicator of the effectiveness of this practice

---

[45] This refrain is often voiced publicly by grieving relatives of victims of Taliban terrorism covered by the local media. For example, in May 2008, a schoolteacher in Kunduz gave a public speech condemning suicide bombers as un-Islamic, and he was himself executed by the Taliban for those remarks shortly afterward. See "Afghan Teacher Shot Dead After Condemning Suicide Bombings as Un-Islamic," *Daily Mail*, May 14, 2008. When acid was thrown in the face of a schoolgirl in Kandahar in November 2008, an Afghan government spokesperson condemned the action as un-Islamic. See "Taliban Blamed for Acid Attack on Afghan Schoolgirls," Associated Press, November 14, 2008. However, the most-compelling condemnation of Taliban terrorism as being un-Islamic was made in June 2009 by a senior Islamic cleric at the Deobandi madrassa in India, which is widely considered one of the ideological pillars that launched the original Taliban movement:

> In an interview with a correspondent of the BBC Urdu Service, the rector and the head of faculty of Darul Uloom (Waqf) Deoband said attacks by "vigilantes" in which innocent people died was not jihad but "individual zulm (oppression)." Seen in this light, attacks on shrines, barber shops and educational institutions were all un-Islamic. Maulana Saalim Qasimi went to the extent of characterising the Taliban regime in Afghanistan, which was ousted by the US forces in 2001, as "un-Islamic." He said the Taliban did not comprehend fully the tenets of Islam even though much was made of their "Islamic government." ("Deoband Ulema Term All Taliban Actions Un-Islamic," *Dawn*, June 20, 2009)

is that the Taliban have assassinated various anti-Taliban mullahs who have stood up against the insurgency with and without government backing. In addition, various USMIL personnel who have returned from the field say that the most-compelling anti-Taliban propaganda is precisely that which labels them and their behavior as un-Islamic. Nonetheless, I conclude that such Islam-oriented messages delivered openly by USMIL mechanisms tend not to be credible and might even be hurtful to coalition efforts. This goes back to one of the key points in the criteria for assessing effectiveness: How credible is the messenger or means of dissemination of the message? The point is not the content of the message but who delivers it and how.

This is not to argue that USMIL personnel should shy away from Islam and Islamic institutions. On the contrary, they are very influential platforms to make a point. The recommendation is that, as a general rule, a non-Muslim should never make judgments in public about what is Islamic or un-Islamic and should not comment on Islamic law or religious doctrine.

Taliban Injustices and Atrocities

An often-repeated staple of PSYOP print-media propaganda showed an actual photograph of Taliban religious police beating women, as shown in Figure 3.14.

Assessment of Effectiveness. Public whippings of women and men did cause considerable resentment among Afghans against the Taliban, and this is an effective theme to this day. However, as noted in the previous section, care must be taken in using Islam directly in USMIL propaganda. Muslim theologians have been arguing among themselves for centuries about what constitutes correct interpretation of the Koran. Moderate, cosmopolitan Muslims might argue that the Koran contains no specific dress code at all, only the injunction to dress modestly, and cite the example of the wives of the prophet who covered their heads. However, a Pashtun might argue vociferously that the Koran specifically mandates the wearing of the burqa. He or she would be wrong in this respect, for the simple reason that the vast majority of Muslim women throughout the world do not wear the burqa, but it would not ameliorate the intensity of the belief about what the Koran

Figure 3.14
Leaflet Showing Taliban Abuse of Women

SOURCE: Reverse of leaflet AFD24 as presented by Friedman, undated (a).
NOTE: Friedman translates the text: "Is this the future you want for your [sic]
women and children?"
RAND MG1060-3.14

says. Given these long-standing internal debates, many Muslims might react with irritation or hostility to any propaganda produced by unbelievers that refers to the Koran or to Islamic beliefs and practices. The unbelievers are seen as not having the authority to use these religious texts for their political or counterinsurgency purposes.[46]

Leaflets Threatening Specific Taliban Leaders

Figure 3.15 reproduces two sides of one leaflet dropped with the intent of sowing fear among Taliban leaders and convincing the population that the Taliban regime was doomed. From left to right on the first illustration, they are Mullah Wakil Ahmad Mutawakil (former Taliban foreign minister), Osama bin Laden, Jalaludin Haqqani (former Taliban minister of borders and tribal affairs), and an unidentified Taliban. The leaflet portrays them as enemies of Afghanistan whose reign of terror will soon come to an end (the corpse-like depictions in the second panel).

---

[46] Author interviews with Afghan mullahs, Pashtun community leaders, and Pashtuns who have assisted U.S. forces in disseminating propaganda.

Figure 3.15
Leaflet Showing the Impending-Demise Theme

SOURCE: Front and back of leaflet AFD56B as presented by Friedman, undated (a).
NOTE: The first three men depicted are, from left, former Taliban foreign minister
Mullah Wakil Ahmad Mutawakil; Osama bin Laden; and Jalaludin Haqqani, former
Taliban minister of borders and tribal affairs. The fourth man is not identified.
Friedman offers this translation: "The Taliban reign of fear [on the front] is about
to end! [on the back]" About two months after this leaflet was disseminated,
Mutawakil voluntarily turned himself in to U.S. forces, seeking to make peace and
serve as facilitator for reconciling other Taliban leaders with U.S. forces and the
new Afghan regime. For that peace initiative, he was imprisoned for two years.
RAND MG1060-3.15

Assessment of Effectiveness. This PSYOP product raises fundamental questions about the contrasting perceptions of the personnel producing the leaflet and their target audience. In the United States, and the West in general, most people have a good idea of what their leaders look like because they see their images frequently on television or in newspapers and magazines. The U.S. public even has a good idea of what enemy leaders look like, because their pictures are displayed in the media. However, in Afghanistan, at the time these leaflets were disseminated, there was virtually no television. Newspapers and magazines were scarce. This continues to be the situation in many rural areas today where the U.S. military operates. Compounding the lack of images of Taliban leaders is the fact that some of them do (or did) not like to be photographed. Mullah Mohammed Omar, the Commander of the Faithful, was notorious in this respect. In contrast, Mullah Mutawakil was photographed many times for interviews with foreign journalists, but these pictures generally appeared in the foreign media and were not available to the great majority of Afghans.

The bottom line is that the most of the target audience probably did not know what their own national leaders looked like, let alone Osama bin Laden. To the illiterate eyes of most of the target audience, the images of Taliban and al-Qai'da leaders in this leaflet might have been seen as just ordinary Afghans wearing turbans. This would have been reinforced by the inclusion of the unknown Taliban at the far right. When the images of these ordinary Afghans are then turned into skulls in the leaflet, the impression could well have been that the U.S. military was threatening death to all Afghans, as opposed to the specific leaders pictured on the leaflet, unrecognizable as leaders to the target audience.

Ironically, Mullah Mutawakil was known in Taliban circles as a dissident, so, of all the pictures of leaders that could have been picked to symbolize the regime, he was probably one of the most inappropriate. Furthermore, including Osama bin Laden in a line-up of Taliban leaders would have been contrary to the views of most Pashtuns, who did not associate this Arab foreigner with their own government. If the intent of the leaflet was to force that association, it probably failed.

According to the criteria of credibility and cultural and historical perspective, this leaflet probably was ineffective in making its point.

Anti-Taliban Leaflets on Terrorist Training Camps

Another anti-Taliban leaflet contained the message, "Do you enjoy being ruled by the Taliban? Are you proud to live a life of fear? Are you happy to see the place your family has owned for generations a terrorist training site?"[47]

Assessment of Effectiveness. Al-Qai'da's training sites might very well have occupied land some family had owned for generations, but this would not have been such a widespread practice as to justify it as a theme in a mass-media campaign. The overall number of al-Qai'da training sites was relatively few, and not that many landowners would have been affected. Having said this, most Afghans indeed had become fed up with Taliban rule. This withdrawal of public support was evident in the rapid collapse of their government when attacked by U.S. and Afghan forces in 2001. More than nine years later, most Afghans still do not want a return of Taliban rule, but local-level polling in Pashtun areas suggests that the Taliban are not seen as mortal enemies of the Afghan people. The ICOS poll indicates the belief that Taliban recruitment is increasing locally, Taliban membership affords higher status, and the Taliban's leaders should be reconciled and given a government position (see Figure 3.16).

Monetary Rewards Are Offered for the Capture of al-Qai'da and Taliban Leaders

- credibility: ineffective
- appropriate context: ineffective
- overall rating: ineffective.

---

[47] SGM (ret.) Herbert A. Friedman, "Psychological Operations in Afghanistan," *psywarrior.com*, undated web page (a).

Figure 3.16
Popular Views of Taliban Recruitment and Status

Was the number of Afghans joining the Taliban in the past year higher, lower, or the same as before?

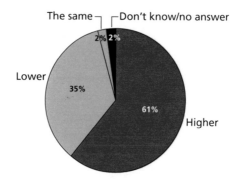

Is membership of the Taliban seen as a high-status social position?

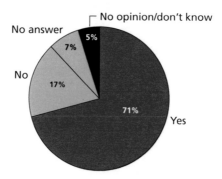

Should Mullah Omar/ the Taliban join the government?

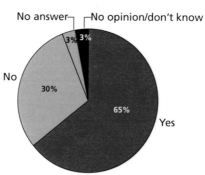

SOURCE: ICOS, 2010.

Reward offers for information on specific leaders are common in counterterrorist and COIN operations, but their results are often disappointing. In the case of Afghanistan, this study found no significant leads generated by this campaign to facilitate the capture of Osama bin Laden or any major Taliban leader. Captures of Taliban leaders generally have been conducted by either Pakistani or Afghan security services, and these must be considered as separate from the PSYOP campaigns aimed at general audiences. Today, this theme is still being disseminated in certain regions, targeting specific Taliban commanders, but it is much less prevalent than in the early period.

Reward for Capture of Terrorist Leaders

Figures 3.17 and 3.18 show examples of the theme of rewards for turning in key terrorist figures. Figure 3.17 is a leaflet that offers U.S. dollars for information leading to bin Laden's arrest and shows him in the orange prison clothes that would later become an international symbol of U.S. imprisonment of Muslim men at Guantanamo, prompting al-

Figure 3.17
Leaflet Advertising a Reward for Osama bin Laden's Capture

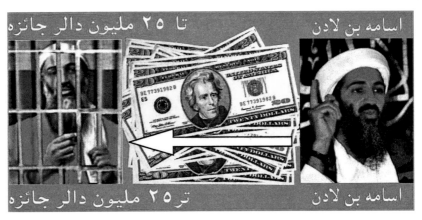

SOURCE: Front of leaflet AFD29n as presented by Friedman, undated (a).
NOTE: Friedman translates the text: "Osama bin Laden" (right) and "$25 million reward" (left).
RAND *MG1060-3.17*

Qai'da militants to dress their captives in similar clothes prior to videotaped beheadings posted on the Internet.

Assessment of Effectiveness. At the time, the Afghani as a currency was just beginning to be implemented, so using a foreign currency was a good idea. Since Americans were making the offer, it also made sense to use U.S. dollars. However, few U.S. dollars circulated in Afghan mountain communities then. Along the Afghan-Pakistan border, the most-used currency was the Pakistani rupee. It might have made more sense to offer the reward in a currency with which the target audience was familiar, as opposed to something exotic. Furthermore, a common deficiency of all reward offers is that they failed to convey to the reader how he or she would be protected from reprisal after taking the life-threatening step to betray a terrorist leader. People who saw these leaflets might have believed that the wealthy Americans would shower them with money if they provided good information on foreign fighters, but they also probably believed that they would not live to enjoy the benefits. Not only was terrorist reprisal a concern, but also local bandits and even the jealousy of their fellow tribe members were formidable psychological obstacles to coming forward. Some of these leaflets, posters, and radio messages urged Afghans with information on terrorist leaders to contact the nearest coalition officials. However, interviews with USMIL personnel who served in Afghanistan during that period indicate that most coalition officials at the local level had received no instructions on how to handle these widely publicized reward programs or what to do with individuals who provided this type of information. This was not a major problem, however, since no leads were generated. The lure of unbelievable wealth could have motivated some Afghans to take the risk and trust their lives to coalition officials, but the bottom line is that probably no one in the target audience knew where Osama bin Laden was hiding. Moreover, if the propaganda cannot convince audiences that they will live to enjoy the huge rewards being offered, it does not make much sense to offer the rewards at all.

Osama bin Laden Matchbook

A second example of the reward theme appears in Figure 3.18. As part of this reward campaign, a green matchbook was created with the bin Laden picture, offering a $25 million reward for information leading to his capture. These matchbooks were distributed all over the Afghan-Pakistan border region where bin Laden was thought to be hiding at the time, but it produced no usable information.

Assessment of Effectiveness. The matchbook uses a problematic approach. The color green is associated in many Muslim countries with Islam. For example, when Muammar Qaddafi published his version of Islamic religious counsel, he titled it *The Green Book*.[48] Thus, Afghan audiences likely saw the printing of bin Laden's image on a green back-

Figure 3.18
Matchbook Cover Offering a Reward for Help Capturing or Prosecuting Osama bin Laden

SOURCE: Matchbook cover circulated in May 2008, as presented by Friedman, undated (a).
NOTE: Friedman translates some of the text: "Contact the nearest U.S. embassy or consulate if you have any information about Osama bin Laden." Inside the matchbook, the text states that the U.S. government wants bin Laden on charges of killing 220 innocents in Kenya and Tanzania in 1998 and that it will pay a reward of up to $5 million for any information leading to bin Laden's arrest or helping to prove the charges against him. The text also reassures the reader of the informer's anonymity and possible relocation out of the country.
RAND *MG1060-3.18*

---

[48] See Muammar Qaddafi, *The Green Book*, Ottawa: Jerusalem International Publishing on behalf of the Green Book World Center for Research and Study, Tripoli, Socialist People's Libyan Arab Jamahiriya, 1983.

ground as portraying a Muslim holy man. The actual picture of bin Laden selected was not unflattering, adding to the positive reaction that most audiences probably had regarding a wanted man they had never seen before in person, much less on extremely scarce television sets or newspapers and magazines. The offer of $25 million has been criticized often among PSYOP specialists as producing cognitive dissonance. It was such a huge sum that it was incomprehensible to the impoverished target audiences living on the equivalent of $2 per day or less. The worst part of this print-media campaign, however, was that it gave no practical means for potential respondents to provide information to U.S. military or civilian officials. Apparently, tribe members living in the remote mountains where bin Laden may have been hiding were expected to make a phone call to U.S. authorities or (although this was not stated explicitly) travel to the nearest U.S. consulate or embassy. Also, the matchbook gave no hint as to what measures would be taken to protect the lives of those who decided to betray the world's most-notorious terrorist in exchange for a huge sum of U.S. dollars.

Monetary Rewards Are Available for Weapons Turned In

- credibility: effective
- appropriate context: effective
- overall rating: mixed.

The offer of monetary rewards for turning in weapons is a message that continues to this day, according to IO and PSYOP personnel interviewed for this study. They indicated that this campaign has produced some good results and is worth maintaining. The current emphasis in some areas is not individual weapons, however, but weapon caches. The MOE in this campaign is that weapon caches and individual weapons continue to be turned in, but not in such amounts as to make a serious dent in Taliban military capabilities.[49] Examples are shown in Figures 3.19 and 3.20.

---

[49] Interviews with IO personnel who have served in Afghanistan.

**Figure 3.19
Billboard Offering Cash Rewards for Turning in Stinger
Missiles**

SOURCE: Friedman, undated (a). Used with permission.
RAND *MG1060-3.19*

Leaflets and Posters Offering Money for Weapons

Assessment of Effectiveness. It is questionable whether the principal target audience for this program, armed Taliban guerrillas, is significantly influenced. The ultimate GIRoA goal is not to collect weapons but to convince insurgent guerrillas to stop fighting. Insufficient evidence exists to make a judgment about whether most people turning in weapons are tribe members needing money or guerrillas who quit. However, the former is more likely: Turning in weapons to the enemy would be a highly offensive act of treason in the eyes of the Taliban, worthy of a death sentence. Given the high stakes, even those Taliban who wish to stop fighting would probably be reluctant to turn in weapons provided by the Taliban. The Taliban have forgiven members who have tired of the fight and gone home, as long as they do not defect to the other side and provide active support against their former comrades. However, participating in a government weapon–turn-in program is a more-serious matter because it deprives the jihad of weapons

Figure 3.20
Poster Offering Cash Rewards for Turning in Weapons

SOURCE: Poster CJTF180-P-AF C031 as presented by Friedman, undated (a).
RAND *MG1060-3.20*

needed for victory. This is a major reason for the assumption that many of those who step forward to turn in individual weapons are not really former Taliban guerrillas but are actually tribe members seeking to make money.

It should be noted that recent information claims that the program offering money for weapons turned in has been very successful. According to an April 3, 2010, press release from the CJTF-82 PA office (PAO) news feed on the DoD reward program,

> So far, in the past three months, Regional Command East has received more than 560 information tips, of which 99 reports led to weapons caches and 18 reports [led] to the capture of key insurgent leaders.

Through this program, Afghans within the Regional Command East area have helped locate more than 1,500 weapons caches and materials in the past three months.

Found weapons caches range from rocket propelled grenades, mortars of various sizes, blasting caps, rockets, projectiles, fuses and other explosive-making materials.[50]

The author has not seen compelling data indicating such a high level of success for other regions during other periods of time. Therefore, the overall assessment of effectiveness continues to be mixed.

Support of Local Afghans Is Needed to Eliminate Improvised Explosive Devices

- credibility: effective
- appropriate context: effective
- overall rating: mixed.

In the beginning of the U.S. intervention, IEDs were not a major problem. As their use proliferated after 2004, so did the PSYOP campaign against them. Today, IEDs are a major PSYOP theme (see Figure 3.21). As an example of PSYOP planning that takes place for all themes, Appendix A reproduces a series of slides detailing the planning, implementation, and evaluation of an anti-IED PSYOP campaign. This appears to be an outstanding effort, reproduced here as an example of good planning. Unfortunately, no data are available on the audience reaction to this particular plan, so its effectiveness is unknown. Of special note in the briefing slides are the attention given to target-audience analysis (both enemy forces and civilian population), means of dissemination, use of key communicators, and MOEs. The three main themes are credible:

---

[50] Bagram Media Center, "Regional Command East Boosts Security, with Afghan Participation in DoD's Reward Program," press release, April 4, 2010.

Figure 3.21
Poster Against Improvised
Explosive Devices

SOURCE: CJTF-76, undated.
RAND *MG1060-3.21*

- IEDs cause harm to local Afghans.
- Al-Qai'da and the Taliban are the perpetrators of this terrorism against the civilian population.
- Come forth and provide information on IED locations and identities of individuals suspected of placing them.

Assessment of Effectiveness. In some places, there have been verifiable, positive results, with local people volunteering critical information. In other places, the locals remain too afraid of the Taliban to come forward. The key variable here seems to be not the credibility of the USMIL IO and PSYOP but the degree of fear of the Taliban and the credibility of the Taliban threat against collaborators.

U.S. Forces Are Technologically Superior

- credibility: effective
- appropriate context: effective
- overall rating: mixed.

PSYOP targeting the Taliban from the beginning have emphasized the technological superiority of U.S. forces and the ability to find and destroy Taliban units, and even individuals, through eyes in the skies and other technological means that are far superior to anything available to the Taliban.[51] This plays into the traditional Afghan attitude that the United States has incredible technological capabilities, bordering on wizardry.[52] For example, some tribe members believe that the United States can implant beacons in peoples' heads without them knowing it, in order to follow their movements.[53] Sample leaflets are shown in Figures 3.22–3.24.

A PSYOP leaflet aimed at a Taliban (enemy force) target audience indicated that U.S. technological surveillance had located Taliban commander Mullah Mohammed Omar and was watching him.[54]

Assessment of Effectiveness. The abundantly obvious technological superiority of U.S. forces, as dramatized most recently by devastating Predator strikes against Taliban and al-Qai'da leaders or facilitators on the Afghan-Pakistan border, gives great credibility to this theme. However, Taliban counterpropaganda points out that the technological superiority of Soviet forces did not save them from defeat and states that the same thing will happen to the United States because Allah is with the jihadists. Nonetheless, recent decisions of the Taliban leadership, as evidenced by independent news reports and USMIL press releases about enemy engagements, suggest a focus on avoiding casualties in confronting U.S. forces. Although there is no Taliban proclama-

---

[51] See McChrystal, 2009.

[52] Author interviews with a broad spectrum of Afghans, as well as with Americans with long experience in Afghanistan.

[53] Interviews with IO and PSYOP personnel who have served in Afghanistan.

[54] Jon Kelly, "The Secret World of 'Psy-Ops,'" BBC News, June 20, 2008.

Figure 3.22
Leaflet Illustrating U.S. Firepower

SOURCE: Front of leaflet AFD40d as presented by Friedman, undated (a).
NOTE: Friedman translates the text: "Taliban and al Qaeda fighters: We know where you are hiding."
RAND MG1060-3.22

Figure 3.23
Leaflet Showing Taliban as Targets

SOURCE: Reverse of leaflet AFD40d as presented by Friedman, undated (a).
NOTE: Friedman translates the text: "Taliban and al Qaeda fighters: You are our targets."
RAND MG1060-3.23

Figure 3.24
Leaflet Illustrating U.S. Surveillance

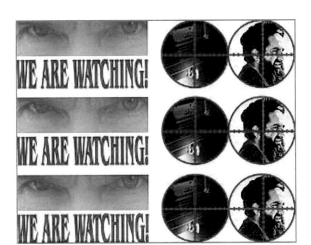

SOURCE: Leaflet presented by Friedman, undated (a).
RAND *MG1060-3.24*

tion to this effect, it is reasonable to assume that U.S. superiority in firepower and technological support plays a role in this more-cautious posture. In their military planning, the Taliban do take into account very carefully the impact of U.S. air superiority and U.S. technological superiority in general. PSYOP thus reinforce a real concern on the part of the enemy.

In terms of dissemination, these leaflets do reach the intended audience and, through pictures and text, convey the intended message. It could well be that an awareness of the overwhelming technological superiority of coalition forces is profoundly demoralizing among Taliban fighters and that PSYOP products emphasizing that point are effective. However, there are little or no data available to corroborate that assessment. At the unclassified level, research for this monograph saw no reporting of significant enemy defections or surrenders due to fear of USMIL technological superiority. Furthermore, there were no unclassified debriefings of captured Taliban available in which this issue was discussed. On the contrary, Taliban surrenders in battle are rare. The Taliban have a reputation for fighting to the death. That being the case, in their minds, enemy technological superiority could be a given, compensated by the righteousness of their cause and their high morale as holy warriors destined for an eternal reward in paradise.[55]

The Afghan Government and Afghan National Security Forces Bring Progress and Security

- credibility: effective (2001–2005) to mixed (2006–2009)
- appropriate context: effective (2001–2005) to mixed (2006–2010)
- overall rating: mixed.

Current PSYOP in Afghanistan place great emphasis on presenting the ANSF, not the U.S. military, as the ultimate guarantors of peace and security. Good governance is another key theme, underscoring the idea that the Afghan government will implement an effective administration in order to counter the spread of an insurgent "shadow government." USMIL guidance to its officers is to minimize the USMIL role in the public media and emphasize ANSF and GIRoA progress.[56] Although the theme of transition has become predominant since early 2010, as far back as 2004, this theme began to be disseminated. Commenting on his 2004–2005 deployment to Kandahar, Colonel Neason stated,

> I think that there was a positive turn . . . because . . . the people began to see things occur, some of it through us, Coalition forces—that is, us and the Afghan army—working with the local government to do [such] things as rebuild mosques and rebuild schools so they can assist with some of the infrastructure rebuilding in Kandahar and the surrounding area. That was, in effect, a way of demonstrating the government's reach because what was important to us as we did things was, emphasized from CJTF on down to us, "We must ensure that an Afghan face is put on all the

---

[55] Interviews with former Taliban leaders, 2009, and review of Taliban propaganda, 2009–2010.

[56] Interviews with IO and PSYOP personnel serving in Afghanistan in 2010.

operations that we're doing. We don't want this to be perceived as [the United States] or the Coalition doing this for them, but in fact try to demonstrate that the government was in fact behind this, and we were just assisting with that effort as they do that," to take and to be empowering to governmental structures."[57]

According to interviews with U.S. military personnel who conducted PSYOP recently in Afghanistan, basic messages include the following:

- The Afghan government is capable of providing security and will defeat the Taliban.
- ANSF are the primary providers of security.
- Joint U.S. and ANSF operations are designed to provide security to Afghans.

An example of the ANSF security role being highlighted by psychological action rather than product can be seen in Figures 3.25 and 3.26 from the 5th Marines Helmand COIN brief, giving excellent guidance to the troops in the field for the conduct of IO and PSYOP.

Besides disseminating the theme that Afghan forces are bringing peace and security to the local population, the 5th Marines COIN brief also emphasized that Afghan forces themselves should distribute these PSYOP products.

Assessment of Effectiveness. Public-opinion polls repeatedly demonstrate that most Afghans view their army with high regard and do expect them to take the lead in the defense of the nation. Also, there is a yearning all over Afghanistan for good governance and a desire for the Afghan national government to fulfill its promises to promote the welfare of the common Afghan. This assessment considers that USMIL efforts to promote a positive image of the Afghan army have been successful, in part because of the audience's receptivity to that theme. However, it has been widely reported that public attitudes about the Afghan National Police (ANP) tend to be much more negative. Given the importance of the police in maintaining law and order at the local

---

[57] Koontz, 2008, p. 363.

Figure 3.25
Slide Emphasizing the Security Role That the Afghan National Security Forces Play

Partner at the lowest level. ANSF enters compounds and buildings first.

SOURCE: 1st Battalion, 5th Marines, undated, slide 2.
RAND *MG1060-3.25*

level, this presents a serious obstacle to the counterinsurgency strategy to promote good governance at the local level. According to the interviews conducted by Strategic Communication Laboratories researchers in the Argahandab and Maywand districts of Kandahar, not only were the ANP seen as being corrupt, abusive, and incompetent; they were also condemned for not being good Muslims.[58]

The civilian side of the government does not fare well either. Reality has fallen short of expectations, and this is having a deleterious impact on popular perceptions of the Afghan government and its ability to take the lead in the fight against the Taliban and the struggle to bring progress. Afghans are openly criticizing corruption that affects their daily lives, from local police officers to diverse local authorities who must be paid bribes to perform their duties. At all levels of society, corruption seems to be flourishing, and ordinary Afghans must navigate an increasingly unjust system in order to make a basic livelihood.

---

[58] Strategic Communication Laboratories, "Perceptions of the Afghan National Police (ANP) in Arghandab and Maywand Districts, Kandahar Province, Afghanistan," London, undated.

**Figure 3.26
Slide Emphasizing a Focus on Afghan Forces and
Government**

Make IO flyers
all about the
ANSF and GIRoA.

SOURCE: 1st Battalion, 5th Marines, undated, slide 8.
RAND *MG1060-3.26*

This situation undercuts the USMIL campaign to put the Afghan government at the head of the struggle for a better Afghan nation.[59]

Some analysts argue that disillusionment with the Afghan government is one of the factors fueling the insurgency:

> The second factor for the [Taliban's] comeback was the increasingly bad governance of the new Karzai administration in which so many Afghans had put their hope and votes. . . . [M]any if not most insurgents are motivated by their rejection of and exclusion by corrupt local government. . . . This is particularly true in most provinces of the South. Here, initially broad tribal coalitions had supported the new administration led by [Hamid] Karzai (himself a Southern Pashtun from the Popalzai tribe). These coalitions

---

[59] Interview with Afghan and foreign journalists in Afghanistan, staff from Afghan think tanks and research institutes, and Pashtun tribal leaders.

were later broken by local strongmen [who] increasingly monopolised power in the name of certain tribes while others were pushed out. A number of those strongmen are either members of the Karzai family, linked to it tribally or through other personal relationship. . . . They also often relied on Western military support (whose mandate was to strengthen the central government and its local representatives) when suppressing protests and resistance—in particular when they were able to label their opponents as [Taliban]. . . . As a result, in Kandahar, Helmand and Farah the Durrani tribal confederation disintegrated into polarised factions.[60]

Given this situation, PSYOP extolling the virtues of the national government are likely to be ignored by at least part of the target audiences. In addition, some data suggest a basic lack of confidence in the Afghan government's ability to provide security and good governance once U.S. and coalition forces withdraw, as suggested in the ICOS poll (see Figure 3.27).

Democracy Benefits Afghanistan, and All Afghans Should Vote

- credibility: effective (2001–2005) to mixed (2006–2010)
- appropriate context: effective (2001–2005) to mixed (2006–2010)
- overall rating: effective (2001–2005) to mixed (2006–2010).

IO and PSYOP regarding this basic theme revolved around two events: the 2004 and 2009 presidential elections (see Figure 3.28). Although Karzai won each time, the differences in terms of IO were profound. In 2004, the Afghan people embraced the idea of democracy and, despite Taliban threats, went to the polls in record numbers. Foreign governments and nongovernmental organizations (NGOs) were closely involved in running the elections, and all pronounced them to be fair. Domestically and internationally, Afghanistan seemed to be on

---

[60] Ruttig, 2009, p. 6.

Figure 3.27
Popular View on the Return of the Taliban

In the areas that have been cleared of the Taliban, if NATO leaves and it is just the Afghanistan government in control, will the Taliban return?

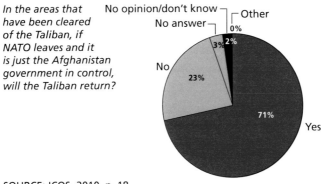

SOURCE: ICOS, 2010, p. 18.
RAND *MG1060-3.27*

Figure 3.28
Leaflets Supporting the 2004 Afghan Presidential Elections

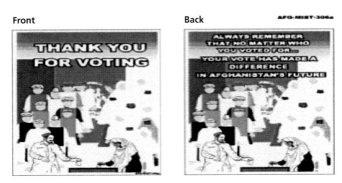

SOURCE: CJTF-76, undated.
RAND *MG1060-3.28*

the road to success. The IO success in supporting the 2004 presidential elections is illustrated in an interview with COL David Lamm, who served as Chief of Staff, Combined Forces Command–Afghanistan from July 2004 to July 2005. He recalled the marching orders prior to the elections in the following manner:

Get out there and do things! Don't do anything immoral or illegal. Other than that, you know the big picture. We've got to have a successful election in October, and whatever that means in your lane—if it means getting a hospital built somewhere, if it empowers the government, if it means getting police training, then that's what you got to do.[61]

These efforts were rewarded:

We had a good election. . . . I mean everything went without a hitch. . . . Actually it was pretty comical to watch the international media. They came to Kabul expecting a freight train wreck and when it didn't happen, they all left town.[62]

In stark contrast, the 2009 elections can be considered a political and psychological failure in terms of bolstering the legitimacy of the Afghan government and promoting the democratic process. As in the cases of air strikes and night raids described earlier in this monograph, the credibility of USMIL IO and PSYOP among Afghan audiences was undercut severely by conflicting statements made by GIRoA and U.S. officials and foreign observers. Whereas foreigners in 2009 and 2010 publicly accused Afghan officials of massive electoral fraud, President Karzai himself accused the foreigners of being the perpetrators of the fraud. In his April 1, 2010, speech to members of the Afghan parliament, President Karzai added,

We have our own national interest in the country. . . . What the foreigners want, and what our national interest is, we have to balance those. If not, our national interests are undermined.[63]

Obviously, these conflicting statements among allies vitiated the potential positive impact of the democracy theme. However, although

---

[61] Koontz, 2008, p. 141.

[62] Koontz, 2008, p. 141.

[63] Joshua Partlow and Scott Wilson, "Karzai Rails Against Foreign Presence, Accuses West of Engineering Voter Fraud," *Washington Post*, April 2, 2010.

the Western media and foreign governments were highly critical of the way in which the 2009 presidential elections were handled, it should not be assumed that ordinary Afghans were as outraged at the foreigners. Predictions of street protests and violence did not materialize. It could well be that the Afghan population became resigned to another round of traditional autocratic government and expected all along that President Karzai would stay in power by hook or by crook.[64]

For additional details on the success of the 2004 election campaign plan, see Appendix B.

Assessment of Effectiveness. In all three criteria, the judgment of this monograph is that the 2004 PSYOP campaign seemed to be effective in large part because the Afghan people were very receptive to the messages. Appendix B reproduces briefing slides outlining the plan of action to support the 2004 election, with excellent data on means of dissemination, specific PSYOP activities before and after the voting, themes and messages, and target audiences.

Regarding the political debacle of the 2009 elections, one should take care to avoid being swayed by the international media bandwagon referred to earlier against the Afghan management of the voting process. An experienced election observer was interviewed for this study, and he stated that what he saw in Helmand province was positive. According to him, polling stations were well run and there was an air of enthusiasm, especially among the women who came out to vote. Some polling stations were run by women, and these were by the far the most efficient and correct.[65] Likewise, IO personnel interviewed at RC East stated unequivocally that, in their areas, people showed up in large numbers and voting went on without a hitch. In contrast to those positive observations, however, there are many more to the contrary: that voting was clearly rigged and that many people stayed away due to Taliban threats. Because of the conflicting reporting, this monograph

---

[64] Typical of the abundant international reporting on the 2009 Afghan presidential electoral are the following: Ghaith Abdul-Ahad, "New Evidence of Widespread Fraud in Afghanistan Election Uncovered," *Guardian*, September 19, 2009; Peter W. Galbraith, "What I Saw at the Afghan Election," *Washington Post*, October 4, 2009; and "U.N. Official Admits Afghan Vote Fraud," *CNN World*, October 11, 2009.

[65] Interview with Norman L. Olsen, official election observer in 2009, 2010.

concludes that the results of the USMIL efforts to support the elections were mixed.

In contrast to 2004, PSYOP support for the 2009 presidential elections emphasized Afghan control of their own elections, with the U.S. role in promoting them publicly downplayed. Because the specific USMIL actions taken to support the 2009 elections are classified, it is difficult to comment on their effectiveness. The basic themes of this campaign were that the presidential elections were good for Afghanistan and that it would be safe to vote. The Taliban clearly mounted a counterpropaganda campaign, arguing that the elections were bad, constituting an ill-fated attempt by a discredited puppet government to gain legitimacy. Moreover, they threatened publicly that those who voted would be punished, specifically warning to cut off the fingers marked with blue ink in the voting process.[66]

Given the publicized election results in 2009, it is evident that voter turnout was low in certain areas, partly out of fear and partly out of lack of confidence that voting would make things any better. The enthusiasm for the 2004 elections was replaced by apathy. It seems that PSYOP failed to convince many Afghans that the elections were safe and merited participation. In Helmand province, for example, election officials say that the voter turnout was less than 10 percent.[67]

It should also be pointed out that there are objections to the way in which democracy has been promoted in general in Afghanistan, even at the beginning, when there was more popular enthusiasm evident. According to some analysts, the seeds of the current crisis of democracy were sown in 2001 in Bonn:

> [I]n the political sphere, a distinct sense of occupation slowly grew amongst Afghans because of the anything but "light" political footprint of the international actors, led by the US. External interference at critical junctions of the political process took

---

[66] Interviews with IO personnel who have served in Afghanistan; review of Taliban propaganda.

[67] Farhan Bokhari, "U.S. Envoy: Taliban Can't Stop Afghan Elections," *CBS News*, August 17, 2009; Farhan Bokhari, "After Attack, Pakistan Confronts Challenge of Burqa-Clad Bombers," *World Watch*, December 26, 2010.

the institution-building process out of Afghan hands, created a group of "most favoured" Afghans . . . and in general discredited democracy as a political option in the eyes of the Afghan public. This included the remote-control induction of Karzai in Bonn, the ousting of the late King and others as Karzai challengers during the 2002 Loya Jirga, arm-twisting in favor of a presidential system . . . , the unconditioned political integration of all [jihadi] leaders and warlords (except Hekmatyar) . . . , the sidelining and neglect of liberal, democratic and civil society forces and political parties. . . . The resulting disenchantment developed into widespread anti-Westernism . . . as a hardening of anti-domination and -manipulation feelings.[68]

If this assessment is correct, it suggests that themes and messages urging participation in the democratic system installed in Afghanistan might not find receptive audiences.

Additional Themes

This monograph contains only the major themes, of which there are many variations. Minor themes identified in the research, such as support for disarmament programs, better security on the Afghan-Pakistan border, and Taliban reconciliation, have not been covered for the sake of brevity. It should be noted that PSYOP campaigns often do not focus on a single theme but might have several woven together in a mutually reinforcing manner. This appears to be effective. For example, during the week of February 19–25, 2005, three main themes were disseminated by the Operation Enduring Freedom CJTF-76. These were the allegiance program, Afghan support for coalition operations, and Afghan successes. Within those main themes, variations of key themes discussed earlier in this chapter of the monograph are evident, including "the Taliban and al-Qai'da are enemies of the Afghan people," "Coalition forces bring peace and progress," and "Afghan government

---

[68] Ruttig, 2009, p. 7.

and ANSF bring peace and progress." Following are the specific messages developing the three main CJTF themes for that week in 2005:

- allegiance program
  - Now is the time for Taliban and Hezb-i-Islami-Gulbuddin Hekmatyar faction (HiG) fighters to rejoin their families, stop following the orders of cowardly leaders hiding outside Afghanistan, and work with the people of Afghanistan to create a better country for all Afghans.
  - A part in Afghanistan's future is available to all the children of this great country. Many Taliban and HiG leaders are already in negotiations with GIRoA to rejoin Afghan society and leave the horrors of the past behind.
  - The United States will continue to target and hunt Taliban, HiG, and al-Qai'da criminals who have no honor and attack the innocent people of Afghanistan.
- Afghan support for coalition operations
  - Coalition forces are working closely with GIRoA, the United Nations, and relief organizations to provide humanitarian relief from the effects of this winter for the Afghan people. The United States is committed to using all of its assets and forces to relieve the suffering of the Afghan people wherever possible.
  - It is the national and holy duty of all Afghans to assist coalition forces in removing threats to a peaceful Afghanistan. Report enemies of Afghanistan and their indiscriminate weapons to coalition forces.
  - By working closely with the leaders and people of Afghanistan, the United States will continue to provide a safe and secure environment in which peace and democracy can prosper.
- Afghan successes
  - Afghanistan has emerged from the terrors of the Taliban as a successful nation. The coalition will continue to support the people of Afghanistan as long as they want its members here as guests.
  - The successes of GIRoA speak for themselves: The Taliban are defeated, the people overwhelmingly chose democracy, the

world community is committed to reconstruction, and the removal of warlords is well under way.[69]

An indicator of how successful PSYOP have been in support of democratic elections is public-opinion polls regarding the leader who won both, Hamid Karzai (see Figure 3.29). Since winning the 2004 election, his "poor" and "fair" ratings in general have increased to the detriment of "good" and "excellent," which saw a steep decline between 2005 and 2009. However, he rebounded briefly in early 2010, perhaps enjoying a Pashtun nationalist backlash against what Pashtuns might perceive as undue foreign criticism of a Pashtun president. Nonetheless, a Pentagon assessment concluded in 2010 that, overall, Karzai was losing support in 2010 among Pashtuns as well as other ethnic groups.[70]

Figure 3.29
Afghan Public Opinion of Hamid Karzai

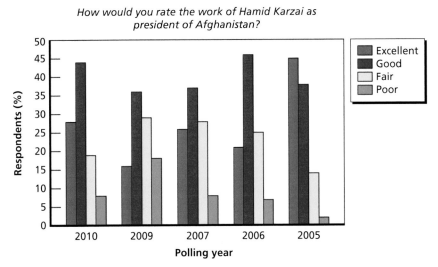

SOURCE: Mynott, 2010.

RAND *MG1060-3.29*

[69] Operation Enduring Freedom, Combined Joint Task Force 76–Afghanistan, "Information Operations: Command Themes Week of 19–25 January 2005," briefing, 2005.

[70] "Afghan Support for Karzai's Government Low: Pentagon Report," Agence France-Presse, April 29, 2010.

CHAPTER FOUR

A Review of the Means of Dissemination in Psychological Operations

Interviews with returning U.S. military IO personnel indicate that the basic means of IO dissemination consist of radio broadcasts, leaflets, press releases, and face-to-face communication with villagers; meetings with local elders are the preferred approach. Each approach is described in this chapter.

Radio

Although the U.S. military continues to sponsor certain radio stations in Afghanistan, most of the radios used are commercial radio stations or Afghan government–controlled radio stations, often owned by provincial governments. In either case, the U.S. military either buys airtime for a public service announcement or provides press releases. Interviews in May 2009 with U.S. military personnel conducting PSYOP in Wardak and Logar provinces indicate that they have become proficient in heading off Taliban propaganda initiatives on civilian casualties. They accomplish this by integrating PSYOP into operational planning and preparing press-release packages before launching a combat operation. Knowing that the Taliban procedure is to seize on actual news to distort it and emphasize any aspect that puts the U.S. military in a bad light, the PSYOP officers covering those provinces have made it a point to beat the Taliban to the punch and get their version of the story on the air before the enemy does. Given the security situation in that region, the commercial radio stations honor requests from the Afghan government, the U.S. military, and the Taliban to put their

communiqués and news releases on the air. It is up to each actor to prepare its contribution and turn it in first.[1]

Assessment of Effectiveness. Since both the U.S. military and the Taliban rely on radio, it should be pointed out that surveys of Afghans show that radio in general is an effective means of delivering messages. Specifically, a 2009 survey by the Asia Foundation shows that Afghans across the country, especially those in districts of interest to the Marines, rely heavily on radio for their information (see Figure 4.1). Table 4.1 summarizes the results of the survey for Helmand province, Kandahar, and across the nation.

For example, the survey shows that 76 percent of Helmand respondents reported that radio was the main source of their informa-

**Figure 4.1
A Pashtun Man with a
Transistor Radio: The Most-
Popular Form of Mass
Communication**

SOURCE: Friedman, undated (a).
Used with permission.
RAND MG1060-4.1

---

[1] Interviews with USMIL officers in Wardak; interviews with IO and PSYOP personnel who have served in Afghanistan.

Table 4.1
Use of Radio by Afghans (%)

Use	Helmand Province	Kandahar	Nationwide
Radio is main source of information	76	46	51
Have confidence in broadcasts	64	60	75
Own a radio	90	87	84
Listen daily	64	47	44

SOURCE: Rennie, Sharma, and Sen, 2008.

tion on what is happening in the country, and 64 percent of Helmand respondents stated that they had confidence in electronic media (such as radio or television). In addition, 46 percent of Kandahar respondents said that they used radio as the main source of their information about the country, and 60 percent of Kandahar respondents expressed confidence in electronic media. Nationwide, 51 percent of respondents claimed to get their information about what is happening in the country from radio, and 75 percent said that they have confidence in electronic media. Furthermore, 84 percent of respondents nationwide actually own a radio, with a roughly similar percentage in Kandahar (87 percent) and a higher percentage in Helmand (90 percent) owning a radio. In Helmand, 63 percent of respondents claimed to listen to the radio every day, while 47 percent of the respondents in Kandahar claimed to do so. It should be mentioned that television is also increasing in importance, but there is not enough information at present to evaluate potential PSYOP utility of that medium.

Figure 4.2 illustrates 5th Marines 2009 guidance for radio dissemination of propaganda, following the pattern established at the beginning of the USMIL intervention.

Figure 4.2
Guidance for Radio Dissemination of Propaganda

Have a team dedicated to running your radio-in-a-box. The enemy doesn't have one and it gives you a distinct advantage in timely dissemination of your message.

Radio-in-a-box works best when the locals have radios to listen.

Have GIRoA officials hand out radios. Have them use the radio to speak to their people.

SOURCE: 1st Battalion, 5th Marines, undated, slides 6, 93, and 101.
RAND *MG1060-4.2*

Leaflets and Posters

The content of specific leaflets, handbills, and posters has been discussed previously. In this section, their general effectiveness as a means of dissemination is evaluated. Leaflets have the advantage that they can be dropped from planes and helicopters into remote areas beyond the reach of USMIL ground patrols (see Figure 4.3). In this respect, only radio has a wider range as a means of mass communication. Leaflets definitely do reach their intended audiences—the air crews can literally see them floating down into the villages. Moreover, PSYOP personnel generally have done a good job of taking into account the low literacy among their target audiences and including pictures or graphics that illustrate the message or theme independently of the text. Whether or not those pictures and graphics were culturally appropriate is another matter. The point here is that these PSYOP products generally have made a good effort to illustrate their messages visually.

The drawback to leaflets is that they are inextricably linked to the U.S. military. Anything dropped out of a plane is associated with

Figure 4.3
U.S. Military Leaflet Air Drop Over Afghanistan

SOURCE: Photo by Warrant Officer 4 Roger M. Gordon, as presented by Friedman, undated (a).
NOTE: Staff Sergeant Dean Penrod drops leaflets north of Kandahar in the spring of 2005.
RAND *MG1060-4.3*

a foreign armed force because the Taliban have never used planes for that purpose. The same goes for material handed out by U.S. patrols or civic-action projects. For some messages and themes, such as "the U.S. military is the friend of the Afghan people," "we are here to help," and "we know where you are," this association of the leaflet as an exclusively USMIL means of communication is an advantage. For other themes, such as "the Taliban (or the terrorists) are un-Islamic," the association could be counterproductive. The same observations hold for posters. Afghans will assume that posters displayed openly in public places were put there by the government or by USMIL forces (unless they are clearly marked by a Taliban logo, since the Taliban also make use of posters). The main advantage of the poster is that it might last longer than a leaflet and thus might not be seen as being as transitory as a leaflet. Both leaflets and posters are seen as government propaganda, however, and this could limit their effectiveness, depending on the message and theme being communicated.[2]

Newspapers and Magazines

The U.S. military continues to use Afghan newspapers and magazines to disseminate message and themes. This effort goes back to the establishment of the newspaper *Peace* when U.S. troops first arrived. *Peace* carried news in Dari, Pashto, and English about Afghanistan and different PSYOP themes. Many of the stories concerned nation building, and, as such, they promoted the peaceful reconstruction of Afghanistan. PSYOP teams gave the newspapers to schools roughly every month as a teaching aid because many schools had no reading material. PSYOP teams also distributed the newspaper to crowds and sometimes within restaurants and shops.[3] Today, the independent print media are more extensive and diverse than they were then, constituting a useful venue.

---

[2]   Author's personal experiences and observations in Afghanistan.

[3]   Friedman, undated (b).

Assessment of Effectiveness. Many observers have questioned the use of print media in Afghanistan because of the high illiteracy rate, especially in rural areas. According to the Economist Intelligence Unit,[4] there are also wide provincial and gender disparities. In 2003, for example, nationwide, 57 percent of men and 86 percent of women above the age of 15 were illiterate; in rural areas, the figures were 63 percent and 90 percent, respectively. Fifteen percent of respondents nationwide claimed to get information from magazines, while only 1 percent of Helmand respondents and 25 percent of Kandahar respondents made this claim. Sixteen percent of respondents nationwide claimed to get information from newspapers, while only 4 percent of Helmand respondents and 28 percent of Kandahar respondents made this claim.[5]

Nonetheless, this monograph maintains that newspapers and magazines are a useful medium for PSYOP message dissemination in Afghanistan because those who can read in a society of high illiteracy often enjoy higher status and are in a position to exert disproportionate influence. Various studies have shown that those attracted to engaging in terrorism tend to be more educated.[6] The Taliban itself was a "student" movement, and Taliban mullahs, by virtue of their ability to read the Koran, enjoy a special position in illiterate rural society. By placing material in newspapers and magazines, the total number of people being reached might be relatively few, but they tend to have much more influence. Moreover, it is not uncommon in Afghanistan for those who can read to orally pass on the content of what they have read, especially current news, to those who cannot read.[7]

---

[4] Economist Intelligence Unit, *Afghanistan: Country Profile*, London, 2008.

[5] Cox, 2006.

[6] See Bruce Hoffman, "Today's Highly Educated Terrorists," *National Interest*, September 15, 2010; Alan B. Krueger and Jitka Malečková, "Education, Poverty and Terrorism: Is There a Causal Connection?" *Journal of Economic Perspectives*, Vol. 17, No. 4, Fall 2003, pp. 119–144; "Exploding Misconceptions: Alleviating Poverty May Not Reduce Terrorism but Could Make It Less Effective," *Economist*, December 16, 2010.

[7] Author's interviews with Afghans involved in the media; personal observations in the field.

Social Networking and the Internet

One set of tools the United States has begun using to build Afghan support for coalition forces is the social-network websites of Facebook and YouTube.[8] Until recently, the United States regularly issued statements denying accusations of misconduct, and release of combat videos was rare. Now, the recent posting of a video underscores a long-held belief within the U.S. military that it needs to be faster and more sophisticated in responding to false allegations. In June 2009, the Associated Press reported that the U.S. military in Afghanistan launched a Facebook page, a YouTube channel, and Twitter feeds as part of a new communication effort:

> Officials said this would help the military reach those who get their information online rather than via printed materials. . . . The effort is primarily to counter Taliban propaganda, which some are saying routinely publicizes false claims about how many U.S. soldiers its forces have killed, or how many civilians might have died in an airstrike. This is the information war which, according to U.S. officials, the military has been losing.[9]

The new effort in Afghanistan is evidently the first in a war zone to try to harness the power of social-networking sites as a primary tool to release information.

Assessment of Effectiveness. Survey data collected in 2008 by the Asia Foundation suggest that it might not be helpful to focus on the Internet as a means of delivering IO and PSYOP messages.[10] According to those data, no respondents (0 percent) in Helmand use the Internet to get information about current news and events, and only 4 percent of Kandahar respondents claimed to get information on current news and events from the Internet. Nationwide, 98 percent of respondents

---

[8]   David Zucchino, "U.S. Fights an Information War in Afghanistan," *Los Angeles Times,* June 11, 2009.

[9]   Dong Ngo, "U.S. Military Joins Twitter, Facebook," *CNET News,* June 1, 2009.

[10]   Ruth Rennie, Sudhindra Sharma, and Pawan Kumar Sen, *Afghanistan in 2008: A Survey of the Afghan People,* Kabul: Asia Foundation, Afghanistan Office, 2008.

claimed they never use it.[11] That said, use of the Internet is growing rapidly in Afghanistan, and its users are often the prime targets for terrorist recruitment: potentially disaffected youths. As with print media, the Internet might be directly reaching very few people in Afghanistan, but these people have enormous potential in terms of being influenced by the Taliban and being swayed to join the radical Islamic cause. The most-compelling argument in favor of exploiting the Internet more fully is that the Taliban itself relies heavily on the Internet to disseminate its ideology and propaganda. As in the case of newspaper readers, it is reasonable to assume that those who use the Internet pass on orally what they see to many others not present during the Internet sessions. However, it should also be taken into consideration that much of the Taliban activity on the Internet might be directed at foreign audiences, as a means of gaining foreign support for the Taliban's cause.

Billboards

The U.S. military currently assists the Afghan government in constructing billboards aimed at undermining support to the insurgency. Typical billboards portray a bold ANSF soldier, a red-eyed terrorist, and a child, with the message that ANSF are protecting Afghans from the insurgents. Billboards are a rarity in Afghanistan and receive special notice. The primary weakness of this method of dissemination is that billboards must be constructed in pro–Afghan government areas where the population is already sympathetic to the message. To mitigate this weakness, billboards are often constructed on roads in pro–Afghan government areas that lead to pro-insurgent areas, thereby exposing at least some of the target audience to the message.[12]

Assessment of Effectiveness. There are neither polls nor interview data on which to base an assessment of target-audience reaction to billboards. However, from anecdotal evidence presented by PSYOP personnel, it seems that billboards reinforce the anti-Taliban sentiments

---

[11]   Rennie, Sharma, and Sen, 2008.

[12]   Interviews with IO personnel who have served in Afghanistan.

of those who are already anti-Taliban and are ineffective in swaying pro-Taliban sectors, as they are seen simply as government propaganda on a big platform. Another possibility to consider is the effectiveness of billboards in areas where the Taliban has established a shadow government. In such cases, they would probably serve to underscore the locals' view of the government's self-delusion and inefficacy.

Face-to-Face Communication

The PSYOP handbook states that face-to-face communication is usually the best means to disseminate messages, and this has been borne out in the Afghan COIN experience. U.S. troops have been engaging in this activity from the beginning, taking advantage of these meetings with village elders, and villagers in general, to reiterate the messages being disseminated through leaflets, posters, and radio broadcasts. Hearing from soldiers directly that they are in their villages to help and provide security and that their only military objective is to defeat terrorists has much more credibility than seeing the same message on a leaflet. There are many examples that could be cited in which U.S. patrols were effective in convincing villagers that they meant no harm and had a genuine interest in their welfare and security. Excellent details on this type of interaction are provided in his 2009 book, *One Tribe at a Time*, in which Special Forces MAJ Jim Gant described his face-to-face meetings in Konar province.[13]

In an interview conducted for this monograph in 2009, a company-level operations officer for Special Forces in Afghanistan makes the following points:

> U.S. forces in Afghanistan conduct hundreds of meetings per day with Afghans in what are termed Key Leader Engagements.

> Platoons are the primary conduit to the population because they have more contact with Afghans than higher elements of

---

[13] MAJ Jim Gant, U.S. Army, *One Tribe at a Time*, Los Angeles, Calif.: Nine Sisters Imports, 2009.

the chain of command. As an example, a standard vehicle patrol would include one or more villages. At each village, the patrol dismounts. Usually Afghans convene on the patrol and present an Afghan man as an elder. The patrol leader speaks to that man, either next to the vehicles or over tea. This exchange allows the patrol leader to gather information and communicate messages to the population while the village expresses concerns. In hostile areas, it is often impossible to find the true leader of the people because he is likely aligned with the Taliban and has reason not to speak to Americans.

Company commanders [also participate in meetings with the] population. They meet with district-level police chiefs, [deputies to the governor], and other local officials. They coordinate their platoons' experiences to try to understand and utilize underlying tribal dynamics.

Battalion commanders have fixed relationships and regularly scheduled meetings with the provincial-level governors, police chiefs, and officials. Periodic shuras allow important figures from the province to communicate with a U.S. military commander capable of allocating substantial resources.

Higher levels in the chain of command also have established ties with Afghans; however, these relationships are more tied to national needs than to specific population groups.

A variety of other groups also conduct regular meetings with the population, including PRTs, agricultural development teams, and human terrain teams.

Key-leader engagements have been prioritized in USMIL's operational plan. For example, in early 2009, after capturing an insurgent, U.S. Special Forces paired with ANSF were required by their chain of command to conduct a multihour meeting with the local population to explain why the man was detained, the crimes that he had committed, the impact that those crimes had on the area, the detention process, and a plan for preventing such arrests in the future.[14]

Assessment of Effectiveness. When face-to-face communication takes place within the context of Pashtun norms of hospitality, with mutual obligations and expectations for guests and hosts, face-to-face communication has been very effective. The Taliban use the population to support nearly all of their operations. Through these conversations, the U.S. military gathers information about these operations and lays groundwork to stop them. These meetings give the opportunity to tap into vast U.S. resources that have been allocated to bolster security, governance, and development. Face-to-face meetings are usually the first step toward bringing these resources to an area. Standard COIN guidance urges that all soldiers should have a PSYOP function and that all of them should see themselves as ambassadors of U.S. goodwill. It is the judgment of this author that visits by U.S. military patrols can also have the opposite effect if accompanied by house searches, breaking down doors, confiscation of weapons, frisking of women, arbitrary detentions of terrorist suspects, and other actions that the Pashtun tribes consider a violation of their honor or of the rules of hospitality.[15]

There have been exchanges between visiting USMIL patrols and village elders shown in televised documentaries that highlight the problem. Often, visiting U.S. soldiers will ask for the location of Taliban guerrillas, or, in a variant to that question, they will ask the villagers directly for help in fighting the Taliban. These questions are extremely stressful and not what polite guests should ask of their hosts for a very simple reason: Answering the question honestly could get the respondent killed. U.S. military visitors usually have no way of knowing whether there is a Taliban informant or sympathizer present at the public meeting. Moreover, even if no Taliban informants are present, gossip of the spoken exchange between a visiting USMIL patrol and the villagers will likely spread, and there is a good possibility that the

---

[14] Interview with a company-level IO officer who served in Afghanistan, 2009.

[15] See 'Abd al-Salam Za'if, *My Life with the Taliban*, Alex Strick van Linschoten and Felix Kuehn, eds., New York: Columbia University Press, 2010.

Taliban will hear it. Once trust has been established (after a series of face-to-face meetings and exchanges of gifts and so forth), the villagers might be willing to discreetly divulge the location of Taliban guerrillas and to commit themselves to fighting the Taliban. However, to expect them to make this life-threatening decision impulsively, in a first-time meeting, in a public setting, is impolite.[16]

Regarding this point, the company-level operations officer interviewed adds,

> Adhering to cultural norms of politeness is a crucial step toward productive communication; however, common-sense propriety must also be observed. In some cases, a platoon leader will arrive in a village, greet the elder, and promptly ask for the location of the Taliban. The elder does not trust the platoon leader yet, so he keeps his information private. The platoon leader may grow frustrated, as he believes [that] the elder is a Taliban sympathizer, and the relationship deteriorates for both parties. Another common mistake is failing to follow through on perceived promises. A village may give information on Taliban activity to USMIL. Giving this information imperils the village, and the village likely expects USMIL action to protect it. The platoon leader who received the information may not understand this relationship. If the Taliban conduct a retribution attack and the village believes that USMIL did not respond sufficiently, the relationship suffers. Likewise, Afghans sometimes make promises in meetings that they have no intention of fulfilling. U.S. forces, which rely on a culture of honor and integrity through honesty, often resent this duplicity.[17]

---

[16] Interview with a company-level IO officer who served in Afghanistan, 2009.

[17] Interview with a company-level IO officer who served in Afghanistan, 2009.

CHAPTER FIVE

Effectiveness in Countering Taliban Propaganda

Although there are notable exceptions, the bulk of the evidence suggests that IO and PSYOP have often failed to counter Taliban propaganda effectively, particularly in the area of civilian casualities and disrespect for Islam. A May 2010 *Afghanistan Digest* article described pessimism among Kandaharis regarding their security situation and negative perception concerning the announced USMIL offensive, suggesting that IO and PSYOP had not achieved their objectives among that target population:

> Caught between the two sides, civilians are hoping to avoid the crossfire. Mohammad Karim, a farmer from Ashgho, said: "The Taliban publicly executed a man in our village by hanging him from a tree and then shooting him. He was accused of passing information to the foreigners. Both sides are creating problems for us and we try to remain neutral." Haji Abdul Haq, a tribal elder from Arghandab district, said people in his area were only interested in avoiding the fight. "The people only want peace and security; they don't care if it's provided by ISAF or the Taliban," he said. A recent public opinion survey in Kandahar conducted for the US army found that despite their efforts to remain above the fray, most of the 1,994 people questioned sympathised with the insurgents' reasons for taking up arms against the government. Some 94% of respondents did not want foreign forces to start a new operation.[1]

---

[1]   *Afghanistan Digest*, May 12, 2010.

Wali Shaaker is a native Pashtun who assisted coalition forces in disseminating propaganda in Afghanistan and worked at the Naval Postgraduate School translating and analyzing Taliban propaganda. Given this unique background, his observations on USMIL IO and PSYOP are worth listening to. He emphasized in an interview for this monograph that the Taliban seek to influence Pashtuns not only through threats but also by calling on Pashtuns' cultural and religious values and pride in their long history of resistance to foreign invasions. Wali made the following points:

> When it comes to design and production of products [that] negate or neutralize the enemy's arguments and accusations, the coalition/U.S. reaction remains far from adequate. It seems that they have simply not been able to generate sufficient responses, in terms of both quality and quantity, to Taliban's intense antigovernment, anti-U.S., and anti-ISAF propaganda.

> Via radio, print media, and even the Internet, the insurgents have been successful in generating and disseminating effective messages, which repeatedly make specific arguments emphasizing certain issues that serve their interests and resonate among the Afghans. In their design of products, generally they exaggerate or invent facts, refer to Koranic verses, as well as narratives of Prophet Muhammad's deeds and sayings, *ahadis* to appeal to religious sentiments of the public. In addition, they constantly allude to the victorious history of Afghanistan, frequently mentioning the defeat of the British, and the demise of the Soviet Communism by the Afghans not too long ago. This is to arouse Afghans' patriotic feelings and direct them against ISAF and the U.S.

> Compared to the Taliban's propaganda, [little] of the literature produced by coalition and U.S. forces appeals to religious and nationalist sentiments of the population in a similar manner.[2]

To improve the appeal of messages disseminated by the U.S. military, Wali proposes seeking assistance from not only the tribal lead-

---

[2]   Interviews with Wali Shaaker, 2009 and 2010.

ers but also the religious elite, writers, scholars, poets, artists (painters, cartoonists), songwriters, stage performers, and school officials in order to achieve greater success in producing and disseminating messages that counter those of the Taliban, and address the main concerns of the public—in other words, build relationships with various segments of the society. He argues for "indirect" means of building rapport with the locals, such as sponsoring sporting events.[3]

Research for this monograph indicates that USMIL IO and PSYOP officers have been seeking aggressively in the past two years to build relationships with the Afghan media. For example, in terms of beating the Taliban to the punch in getting press releases on USMIL operations in a timely manner to local media, PSYOP officers have been successful on Logar and Wardak radio stations in countering Taliban propaganda.[4]

In the past, a contrary situation has been more common, as illustrated by the failed effort to counter Taliban claims concerning U.S. air strikes during the night of May 4, 2009, in the Helmand province village of Granai. Almost immediately, the Taliban propaganda machine claimed that 140 innocent civilians had been killed, a charge repeated by the Afghan government itself. The United States denied the allegations and, over the next couple of weeks, made the following counterclaims by press releases and statements of U.S. military officials:

- Innocent civilians were killed, but far fewer than alleged, and it was the Taliban's fault.
- The Taliban deliberately attacked coalition forces from this village and prevented civilians from leaving in order to maximize civilian casualties for propaganda purposes.
- The Taliban knew which house would be bombed and herded people into it.
- The Taliban were seen transporting dead bodies of civilians in pickup trucks and bringing them to the site (possibly suggesting that the Taliban killed the civilians).

---

[3] Interviews with Wali Shaaker, 2009 and 2010.

[4] Interviews with PSYOP officers in Wardak and Logar, 2009.

At the end of this series of increasingly improbable scenarios, the U.S. military spokespeople finally admitted that the plane that dropped the bombs failed to follow correct procedures and verify that no civilians were in the target zone below. In effect, the Taliban were no longer the main culprits, and the United States military now admitted that it had made a mistake but remained firm in insisting that the actual casualty figure was around 32 civilians as opposed to 140.[5]

In terms of damage limitation, this performance probably hurt the U.S. image. This episode highlights the need to integrate PSYOP and PA functions more closely, a recommendation that many officials oppose as being contrary to doctrine. (See a fuller discussion of this issue in Chapter Six.) After providing a series of inconsistent explanations, the United States wound up admitting fault. Part of the problem is that standard counterpropaganda messages, such as "the enemy uses civilians as human shields," no longer seem credible with Afghan audiences. Apparently, they believe that U.S. and coalition forces simply do not value Afghans' lives and prefer to drop bombs on innocent villagers rather than take casualties themselves. For years, the U.S. government rationale has been that civilian casualties are unavoidable if combat operations against terrorists are to continue in the Afghan theater. The danger associated with that rationale is that the Afghans might infer the corollary that foreign combat operations must cease in order to stop the deaths of innocent civilians.

Various studies have noted the far-reaching political repercussions of the U.S. military's inability to mount an effective counterpropaganda campaign on the issue of civilian casualties. For example, Colonel Main writes,

> The Taliban conducted an effective information campaign, alleging excessive non-combatant [casualties] from coalition airstrikes in Afghanistan. US forces were unable to convince the local

---

[5] See Mark Tran, "US Military Admits Errors in Air Strikes That Killed Scores of Afghan Civilians," *Guardian*, June 3, 2009; Shapoor Saber, Fetrat Zerak, and Abaceen Nasimi, "Soul-Searching Following Farah Tragedy," Institute for War and Peace Reporting, June 5, 2009; and Elisabeth Bumiller and Carlotta Gall, "U.S. Admits Civilians Died in Afghan Raids," *New York Times*, May 7, 2009.

and international media that they did not kill innocent civilians in these airstrikes. That negative message cost the President of Afghanistan popular support. This misperception was reinforced by the local and international media, communicating to the Afghan people, asserting that the US was bombing compounds and killing large numbers of Afghan civilians, without an acceptable reason. The President of Afghanistan losing popular support attempted to limit the coalition use of airpower in Afghanistan.[6]

This effort to limit airpower ultimately proved successful. Since July 2009, new directives were issued concerning use of air strikes. Concerning the new restrictions, Colonel Main makes the following observation about the utility of close air support:

> Airpower in support of small units has allowed coalition forces to greatly expand the areas in support of the Afghanistan Government. The inability of PSYOP or Strategic [Communications] to address this perception of excessive [casualties] has restricted one of the most effective kinetic tools available to the coalition.[7]

On the other hand, as a direct result of the new restrictions, civilian casualties due to air strikes have decreased significantly. This statistic corroborates in the minds of many Afghans the long-standing belief that the United States had the power for the past nine years to curtail civilian casualties. The DoD *Report on Progress Toward Security and Stability in Afghanistan*, submitted to Congress on May 28, 2010, states,

> Civilian casualties (CIVCAS) is a strategic issue that will impact the success and progress of the U.S. and international community in Afghanistan. Minimizing the number and magnitude of CIVCAS incidents is critically important, as is the need to effectively manage the consequences of such incidents when they do occur. The insurgents are responsible for 80% of CIVCAS. However, insurgents can exploit and manipulate CIVCAS events to

---

[6] Main, 2009, p. 5.

[7] Main, 2009, p. 5.

their advantage, while the U.S. and international forces are held accountable by the Afghan population for all incidents where there are CIVCAS.[8]

That last observation is further corroboration that the U.S. military is not being effective in countering Taliban propaganda.

Besides public-opinion survey data cited earlier in this monograph, showing a decline in support for the Afghan government and its foreign patrons, other evidence suggests a deepening of social conflict in Afghanistan. This situation can obstruct U.S. interests and contradict PSYOP themes of harmony and progress in a democratic, multiethnic Afghanistan. For example, in his most-recent publication, Gilles Dorronsoro, who has conducted extensive research in Afghanistan for the Carnegie Endowment for International Peace, describes the current situation in very negative terms:

> In all my visits to Afghanistan since 1988, I have never seen as high a level of distrust and hostility between Pashtuns and other ethnic groups as I witnessed during April 2009. . . . As a result of the changing balance of power between the different groups, the Pashtuns have been discriminated against in the North of the country, where they are a minority. The local administration excludes them and exactions are frequent. Complaints to the Ministry of Interior or Tribes produce few results, leaving Pashtuns feeling further victimized.
>
> Second, communal and sectarian conflicts, which were essentially local in scale, now resonate throughout the country. In particular, the Afghan media [have] played a major role in expanding the geographical scope of ethnic and sectarian conflict. Far from promoting understanding between sectarian or ethnic groups, media outlets have actively fueled resentment in the last few years. Major political competitors own TV and radio channels and use them for mobilization purposes.[9]

---

[8]   DoD, 2010b, p. 43.

[9]   Gilles Dorronsoro, *The Taliban's Winning Strategy in Afghanistan*, Washington, D.C.: Carnegie Endowment for International Peace, 2009, p. 13.

In interviews conducted for this monograph with Pashtun tribal leaders and former Taliban members in Afghanistan between April and May 2009, the same, unprecedented bitterness among all those interviewed was apparent. Throughout the Pashtun belt of southern and eastern Afghanistan, it was repeated that intense resentment has developed against U.S. and coalition forces because of their tactics. These tactics are considered violations of the Pashtunwali (code of the Pashtun). The most-common accusations include those listed here:

- searching private homes, breaking down doors, terrifying and humiliating Pashtun families
- nighttime raids, entering bedrooms and women's quarters at night, sometimes resulting in shootings of innocent villagers who try to fend off these assaults
- frisking of women in their homes and in public places
- arbitrary and indefinite detentions of local villagers taken to U.S. military bases outside the jurisdiction of Afghan national law, Islamic law, and tribal law
- frequent killings of innocent civilians during air strikes and combat operations
- disarmament policies that leave villagers vulnerable to bandits and insurgents.

Concerning these types of complaints against USMIL and NATO actions, Ensign Bebber's earlier 2008 survey in Khost indicated the following:

> There seemed to be an increase in complaints about Coalition Forces conducting searches—especially at night—and the growth in civilian casualties. Locals complained that only Afghan security forces should be permitted to conduct searches, and that they could be conducted during the day. They wondered why the Coalition could not just surround a house at night to prevent anyone from escaping and then wait until morning so the entire village could see who was being arrested and why. It should be noted that villagers said they had no problem with arresting those who aid the Taliban in any way. They also said that if vil-

lages permitted insurgents to use their village to stage attacks on Afghan and Coalition Forces, it was appropriate for the Coalition to respond, even if civilians are killed. This follows the *Pashtun-wali* tradition of attacking those who permit their homes to be used as a base to conduct attacks on others.[10]

Various published studies show how Taliban propaganda skillfully exploits complaints about U.S. and coalition tactics.[11] Evidently, USMIL PSYOP have, for years, been largely ineffective in countering this form of Taliban propaganda. However, this situation is changing. When General McChrystal took command in July 2009, he began to issue directives ending or curtailing these tactics. For example, U.S. troops generally do not search homes any more (this being left to Afghan troops) and do not frisk women. Taliban propaganda, of course, continues to publicize the past practices, ignoring the changes that have taken place. To combat this, USMIL IO and PSYOP have implemented new approaches that are proving to be very effective. These include training Afghan journalists and establishing working relationships with them, as well as establishing relationships with respected tribal leaders and Islamic clerics. As a result, the last time the Taliban sought to generate riots over fabricated reports of USMIL desecration of Korans, these local key communicators discredited the charges and were able to maintain calm in their communities.[12] This is an IO success achieved through Afghan intermediaries.

Moreover, there are an increasing number of examples in which IO have been integrated into operational planning and the U.S. military has preempted Taliban propaganda initiatives. Commander LeGree provides a compelling case study of IO and PSYOP effectiveness in the Afghan theater:

---

[10] See Bebber, 2009.

[11] Dawood Azami, "Taliban Slick Propaganda Confronts US," *BBC News*, August 3, 2009.

[12] Oleg Svet, "Fighting for a Narrative: A Campaign Assessment of the US-Led Coalition's Psychological and Information Operations in Afghanistan," *Small Wars Journal*, September 12, 2010.

In the summer of 2007, a platoon-plus element from Able Company, 2d Battalion, 503d Infantry, 173d Airborne, conducted an airborne insertion into Sangar Valley in Kunar province. The insertion was to be of limited scope and duration, to show a presence and conduct shaping operations. The battle for the minds of the people was at a critical stage in this valley, an area historically supportive of the Taliban, but showing signs of opening up to the Afghan government. . . .

Realizing the ferocity and scope of the operation, the PRT and battalion immediately mobilized concurrent coordinated real-time IO with the provincial governor. We sought to beat the Taliban to the news cycle and highlight the atrocities underway. The chaos of the battlefield meant it would be days until we could evaluate the final details, but we had no trepidation about telling the story as it unfolded. . . . We felt [that] it was better for the people to hear about the battle immediately and from a credible Afghan source. The PRT made quick contact with the Ministry of Defense, and the ANA [Afghan National Army] deputy corps commander flew to the provincial capital of Asadabad within two hours. We immediately held a radio and television press conference complete with maps and relevant details of the engagement, provided constant press updates as the battle unfolded, and maintained a *credible* public dialogue.

By acting *inside* the Taliban's news cycle, we put the insurgents on the defensive. They lost the advantage of initiating a story. . . . We did not forget the battle for the minds of the people during the heat of the lethal battle. Our efforts to connect the people to their government were successful, despite the worst of circumstances, and the credibility of the government as a voice of reason and authority in a time of crisis improved. . . .

In fact, the local IO effort had a wider effect. The press conferences received national attention, and the story was one of several accounts of the Taliban intentionally targeting civilians. This damaged the Taliban's credibility. . . . Although there were casualties, truth was not one of them, and trust in government was reinforced.[13]

Other specific examples of how USMIL and NATO forces dealt effectively with Taliban propaganda challenges and built good relations with local communities to undercut the insurgency are included in the excellent 2010 compilation by the Center for Naval Analyses (CNA) titled *Counterinsurgency on the Ground in Afghanistan: How Different Units Adapted to Local Conditions.*[14] That study highlights the fundamental nature of PSYOP as a multifaceted activity, illustrating the concept articulated earlier in this monograph that every infantryman is a de facto PSYOP actor in the field.

---

[13] LeGree, 2010, p. 25.

[14] Jerry Meyerle, Megan Katt, and Jim Gavrilies, *Counterinsurgency on the Ground in Afghanistan: How Different Units Adapted to Local Conditions*, Center for Naval Analyses, November 2010.

CHAPTER SIX

Organizational Problems Affecting Information Operations and Psychological Operations

Interviews with USMIL personnel returning from the field, a review of OEF conducted by Major Cox at the U.S. Army Command and General Staff College, and other USMIL and academic studies reveal a set of interrelated problems summarized as follows:

- lack of standardized IO and PSYOP integration with operations
- long response times and coordination-process delays
- ineffective interface between IO and PSYOP
- isolation of IO officers
- conflicting IO, PSYOP, and PA functions
- failure to exploit the informal, oral Afghan communication system
- general lack of MOEs.

It should be emphasized that these problems are not meant to be a universal characterization of all IO and PSYOP and that their applicability varies between units and over time. A lot has happened since Major Cox wrote his review, even during the time it has taken to do the research for this study. During the past year, for example, some field commanders have placed great emphasis on integrating IO and PSYOP into their operational planning, as mentioned in the example of the Battle of Sangar in Chapter Five. Systematic efforts to improve coordination are being initiated. Nonetheless, it is useful to summarize the critiques that have been made to use them as a frame of reference for making current evaluations.

Lack of Integration of Information Operations with Unit Operations

In a monograph on IO in OEF, Major Cox explained that, from their conception, IO were always intended to be integrated into a unit's operations.[1] But because doctrine provided commanders so little guidance on how to integrate IO effectively, commanders had to figure it out for themselves. In fact, Major Cox concludes that problems with integrating IO into operations have stemmed directly from a lack of guidance in doctrine. If a commander considered it important to integrate IO into operations, he expressed the importance of IO in his intent, and only then did integrating IO become important to his staff and subordinate commanders. Major Cox also emphasized that commanders had difficulty grasping how the media operate and how to use media to their advantage. His assessment elaborates further: Most failures to integrate IO have occurred because the commander did not visualize the complete operational environment. Commanders have often viewed IO only in terms of what can be presented in the media; as such, they have used IO to help spread good news (inform) rather than change a target audience's perceptions (influence), degrade their adversaries' ability to manage perceptions (attack), or even defend the information environment the commander had been trying to create (protect).[2]

Lack of integration of IO and PSYOP with operations was an often-repeated complaint in interviews with officers returning from the field. However, there was also strong contrary evidence that this problem is being addressed. The Marja campaign is an outstanding example of the close integration of IO and PSYOP with operational planning, as alluded to in the quotations from ADM Michael G. Mullen in Chapter One. An illustration of that process can be seen in Figure 6.1, from the 1st Battalion 5th Marines Helmand COIN brief.

---

[1]   Cox, 2006.

[2]   Cox, 2006, p. 3.

Figure 6.1
Promoting Close Integration of Psychological and Information Operations with Overall Operations

Use simple, clear signs to convey your message. They degrade quickly, so have back-ups pre-made.

SOURCE: 1st Battalion, 5th Marines, undated, slide 5.
RAND *MG1060-6.1*

Long Response Times and Coordination-Process Delays

Placing a message on a local radio station, under standard guidance that it should not appear to be USMIL propaganda, requires going up the chain of command and obtaining approval at battalion level, then at brigade level. This can take up to a week. Some messages are time sensitive, particularly if they are keyed to ongoing operations, and such delays can vitiate their effectiveness. A common complaint is that, although everyone pays lip service to integrating IO and operations, the lengthy coordination process and inherent delays mean that the IO element often is ignored in operational planning and execution.

The complaints about the approval process and its inherent delays are typical of PSYOP as well. In *Enlisting Madison Avenue*, the authors write, "The process for approving PSYOP products has been criticized for its lack of timeliness. By the time some products are approved, fast-paced events have too often negated their value."[3] In his PSYOP lessons-learned study, Christopher J. Lamb concludes,

---

[3]   Helmus, Paul, and Glenn, 2007, p. 155.

A dilatory PSYOP product approval process is detrimental to the execution of an effective PSYOP campaign. Before operations begin, a delayed process inhibits PSYOP planning and rehearsal time, while slow approval during an actual campaign can render some military and political products useless, since they may be overcome by events. Unless the approval process is reformed, both at the theater and tactical level, PSYOP effectiveness will be seriously compromised.[4]

Ineffective Interface Between Information Operations and Psychological Operations

As stated in an interview with an IO officer who had served in Afghanistan, because IO officers do not have the resources to produce leaflets or other propaganda products themselves, they must rely on PSYOP officers and their capabilities. However, the PSYOP structure is described as unwieldy and unresponsive: "Requests for PSYOP products do not match any [operational] tempo." Under the current PSYOP coordination system, leaflets that could have a significant effect if produced within 24 hours and distributed immediately on the battlefield can take as long as a month to produce.[5]

This issue goes back to the discussion in Chapter One on overlapping missions and doctrinal disagreements. IO officers complained in interviews that they did not have the resources to produce leaflets and other products. Doctrinally, however, it is not the function of IO officers to produce leaflets in the first place, nor even ask for them. This underscores the doctrinal disagreements that hamper unified action. On this issue, Lamb concludes,

> As part of its recommendation set, the *Information Operations Roadmap* suggested that PSYOP become integrated with broader IO efforts. . . . [A] major problem documented in OEF . . .

---

[4] Lamb, 2005, p. 14.

[5] Interview with IO officer who served in Afghanistan.

lessons learned is that IO planners did not adequately understand PSYOP and thus failed to appreciate [PSYOP's] capabilities or employ them appropriately or effectively. . . .

It is also true that many IO officers were not well trained. . . .[6]

Isolation of Information Operations Officers

According to some interviews, IO officers are physically separated from the operations center (OPCEN) and thus do not have good knowledge of what is going on in the present and what is being planned for the future. The complaint is that physical separation reinforces an operational separation, which negates the mandate to integrate IO with operations.[7] Moreover, given the potential political ramifications of certain media initiatives, specific messages tend to be written, or receive final approval, at the higher levels of the chain of command. A key problem is that distribution tends to take place at that level, which often is not linked to local media at the local level, at which operations take place. In concrete terms, this means that messages released to the media in Kabul frequently do not filter back to the provincial communities, which constitute the main target audiences, because there is little interface between the national capital and local communities in terms of news dissemination.

Interviewees added other observations: Ground troops cannot rely on higher echelons to perform some PSYOP functions. In practice, press releases, radio broadcasts, and relationships with Afghan media are often centralized at the brigade level. As actions occur on the ground, such as the wounding of a civilian or the capture of an insurgent, the U.S. message needs to reach the population quickly and accurately. After a combat action, a report recording the event is compiled at the platoon level then examined by three levels of the chain

[6] Lamb, 2005, p. 15.

[7] Interviews with IO and PSYOP personnel who have served in Afghanistan, 2009 and 2010.

of command before reaching the brigade-level IO officer, who then writes a press release in English. As information about the event on the ground travels up the chain of command, it loses timeliness, context, and clarity. Interviews reveal that subordinate leaders often downplay events in order to avoid involving higher headquarters (HHQ) with their operations. The company commander, who understands the context of what occurred and can integrate the message into other local operational efforts, could more-effectively write the press release and radio broadcast. Company leadership should also expand ties with local media sources rather than relying on the brigade IO officer. By establishing this relationship, local media will know whom to contact to get information regarding local events and U.S. troops. This places significant responsibility in the hands of company leadership who are not trained in IO.[8]

In contrast, some observers comment that it is not practical to have an IO or PSYOP specialist at every company or platoon. Instead, they say, every soldier should be a communication platform:

> Commanders should not underestimate the value of face-to-face activity and using host nation capabilities. The latter underscores the need to think beyond USMIL capabilities and begin to rely more [heavily] on host nation [HN] platforms (ANSF, HN key communicators, HN media) as suggested by the recommendation . . . that company leadership should also expand ties with local media sources. It is very important because it adds credibility [and] improves the product based upon advice, and the medium is more readily consumed by the target audience. There are times that dissemination by the US is important, but typically not as the only method.[9]

---

[8]   Interviews with IO personnel who have served in Afghanistan, 2009 and 2010.

[9]   Friedly, 2010.

Lack of Coordination Between Information Operations and Public Affairs

In the Granai affair discussed earlier, public statements of U.S. commanding officers, as well as U.S. military press releases, were intended to influence Afghan audiences, as well as to inform U.S. audiences at home. This highlights the conflict between IO, PSYOP, and PA functions and the need to integrate those more closely. There is no question that whatever U.S. spokespeople said in public about the deaths at Granai could have a significant psychological effect on target audiences in Afghanistan, as well as throughout the Muslim world. The jihadi narrative is that Islam is under attack, and air strikes killing innocent Muslim civilians are taken as confirmation. In this context, the PA press releases and oral statements, replayed by the local media, were squarely within the PSYOP arena.

In his unpublished *Speed Versus Accuracy* paper, Colonel Scott makes the following points on how to best respond to enemy propaganda claims:

> Speed is important when reporting unfavorable news resulting from the actions of friendly forces. Releasing factual information related to negative events prevents the negative credibility [that] results from allowing the enemy to release the information first. Failure to apply speed in releasing news of negative action gives the appearance of a cover up, a lack of transparency. It enhances the effectiveness of enemy propaganda by allowing [the enemy] to release the information first. The delayed release by friendly forces either becomes an endorsement, or confirms the accuracy of the enemy's information thereby increasing [its] credibility.

> In February 2007, an incident in Afghanistan provided an example of the risk associated with applying speed in response to a crisis event without collecting and confirming the facts and de-conflicting the message within the organization. A suicide bomber attacked a Khost hospital opening ceremony. Different U.S. elements and the local media participating in the ceremony immediately began to disseminate different accounts of the event. After several weeks of attempting to correct the initial misinfor-

mation disseminated, the end result remained unchanged. The local audience perceived the U.S. to have intentionally spread disinformation concerning the event.

Dissemination of inaccurate information affects the "cores of credibility" of integrity, intent, and capability of the organization. Inaccurate information damages the organization's reputation of truthfulness and results in an incongruence between actions and words. It makes the organization look inconsistent and displays a lack of transparency. Disseminating inaccurate information requires retractions and corrections, which in turn make the organization look incompetent. This does not mean speed should be sacrificed to mitigate the risk to credibility.[10]

Achieving the right mix of speed and accuracy as proposed by Colonel Scott presupposes close coordination between IO, PSYOP, and PA. However, the official view of the PA office is that it must provide accurate information without any intent to influence. Official doctrine follows:

(b) PA and PSYOP products should provide a timely flow of information to external and internal audience. Based on policy, PA and PSYOP must be separate and distinct even though they reinforce each other and involve close cooperation and coordination. Each function requires distinct efforts to plan, resource, and execute as part of the commander's operation plan (OPLAN). It is critically important that PA and PSYOP coordinate with each other to maintain credibility with their respective audiences. Therefore, PSYOP representatives should coordinate with command PA offices supporting the joint information bureau and PA representatives present within joint planning organizations such as the joint planning group, operations planning group, or information operations (IO) working group to integrate operational activities while strictly maintaining autonomy.

---

[10] Colonel Jeffrey Scott, *Speed Versus Accuracy: A Zero Sum Game*, unpublished paper, 2010.

(c) PA and PSYOP products must be coordinated and decon-flicted early in the planning process and during execution. Although PA and PSYOP generated information may be differ-ent, they must not contradict one another or their credibility will be lost. Although each has specific audiences, information often will overlap between audiences. This overlap makes deconfliction crucial. Under no circumstances will personnel working in PA functions or activities engage in PSYOP activities. Commanders will establish separate agencies and facilities for PA and PSYOP activities, with PA being the commander's primary contact with the media.[11]

This division of labor is logical and well founded. If PA were seen as a propaganda arm of the military, it would lose credibility and effec-tiveness. Nonetheless, there are situations in which PA and PSYOP have been combined. During the first Gulf War, for example, PA offices reported on plans for an amphibious landing, which was a deception. The Marines were positioned offshore, making every preparation to invade. However, the objective was to tie down part of Saddam Hus-sein's army in that area so that it could not take part in battles where the real thrust was to take place. PA participated in that deception operation. In a different sense, PA statements and press releases regard-ing Granai also had a PSYOP impact. This underscores the problem that, although the enemy has implemented a unified anti-U.S. pro-paganda campaign, the United States subdivides IO, PSYOP, and PA functions, creating discrete units with separate missions. Christopher Paul addressed this controversy in his IO handbook:

> Counterpropaganda features prominently in PSYOP doctrine, but is also part of the public affairs (PA) portfolio. It isn't clear who has the lead. Further, while there is some evidence of coun-terpropaganda activity during contemporary operations from both PA and PSYOP activities, author interviews with PSYOP personnel in 2006 suggested that counterpropaganda was a low priority for a very busy organization and received very little

---

[11] FM 33-1, 1993, p. 1-9.

attention. . . . Counterpropaganda is an area for which IO integration makes sense. Several IO or related components could integrate in this often rapid response area of information warfare. Curiously, though, of the two capabilities tagged in doctrine with responsibility in this area, only one, PSYOP is officially "IO," with PA being a "related" capability.[12]

In September 2004, the Chairman of the Joint Chiefs of Staff, Gen Richard Bowman Myers, issued a policy memorandum to the joint chiefs and commanders of the combatant commands that IO and PA must be separate staff functions. These functions could coordinate, but General Myers cautioned against the intermingling of IO and PA. According to his memorandum, the purpose of IO was "to influence foreign adversary audiences using psychological operations capabilities." For its part, PA should focus on informing the "American public and international audiences in support of combatant commanders' public information needs at all operational levels."[13]

Some policy planners have maintained that the firewall between IO and PA is more urgently needed today than in the past. Until Operation Desert Storm, it was difficult for information disseminated in theater to reach the United States. Now, with the advent of the Internet, satellite phones, computers, and portable short-wave radios, it is possible to pick up such content and relay it to the United States virtually instantaneously. Commanders understand that the U.S. audience and the foreign audience are different. However, given modern technology, it is difficult task to prevent "spillage." One notable example occurred during Operation Desert Storm, during planning for a potential landing of Marine Corps personnel in Kuwait. The PAO was aware that the landing was, in fact, a deception and knew how important the decep-

---

[12] Paul, 2008, p. 67. Approaching this issue from a different perspective, Maj Tadd Sholtis makes various recommendations for dealing with the media and improving coordination among USMIL components in this area. See Maj Tadd Sholtis, "Planning for Legitimacy: A Joint Operational Approach to Public Affairs," *Air and Space Power Journal*, June 8, 2005.

[13] Gen Richard B. Myers, Chairman of the Joint Chiefs of Staff, "Policy on Public Affairs Relationship to Information Operations," memorandum, CM-2077-04, September 24, 2004.

tion was to the commander's battle plan. In his analysis, Major Cox argues that the command should have minimized the PAO's knowledge and involvement in the deception planning. This would have kept the PAO apart from operations, thus allowing the office to maintain its ethical standards and credibility by not being put into a position of deliberately providing false or misleading information.[14]

IO and PSYOP advisors who have served in Afghanistan and Iraq point out another dilemma that arises as the theater IO environment matures. PSYOP teams are generally used to dealing with the local population and the indigenous press. However, as the media landscape matures, these same press members tend to demand treatment and access similar to that extended to the international press and want to be included in the international press briefings. The U.S./coalition saw this in Iraq. In such situations, the PSYOP teams, as an element of IO, may be attempting to reach the indigenous press with one message while the PAO may be attempting to reach that same indigenous press with a different message. To achieve overall IO objectives, this line of thinking argues that there should be closer coordination and deconfliction between IO and PAO messages, as opposed to greater compartmentation.

Also using Iraq as an example, Major Richter argues that the experience resulted in a greater awareness of the need for better integration of PA and IO:

> Like others did in the Balkans and Afghanistan, Colonel Ralph O. Baker, a brigade commander in Iraq, discovered the operational significance of public information and the subsequent need for PA and IO integration. He realized that press releases, whether Iraqi or international, have immediate effects on popular attitudes and can counter enemy propaganda. To assist Baker's IO planning, PA provided him with media analysis on popular perceptions in sector.[15] . . . Despite the contentiousness of the

---

[14] Cox, 2006, pp. 83–85.

[15] COL Ralph O. Baker, "The Decisive Weapon," *Military Review*, May–June 2006, pp. 13–32. (Footnote in original.)

IO-PA issue, most senior military leaders acknowledge the need for effective PA-IO integration.[16]

The 2003 *Information Operations Roadmap* also zeroed in on this problem with its recommendation to "Clarify Lanes in the Road for PSYOP, Public Affairs and Public Diplomacy."[17] It called for greater coordination between DoD PA and other U.S. government agencies—in particular, the State Department Office of Public Diplomacy and Public Affairs—stating that "PSYOP forces and capabilities may be employed in support of public diplomacy," while keeping the main focus on "support to military endeavors in nonpermissive or semipermissive environments (i.e., when adversaries are part of the equation)."[18] Following up on the 2003 road-map recommendation, Oleg Svet's 2010 assessment reviewed the roles of the State Department, DoD, and Central Intelligence Agency (CIA) in this area and concluded, "With diffused authorities, it has been difficult to pursue a comprehensive narrative providing legitimacy for the local government, quickly respond to the Taliban's propaganda, and proactively shape the information environment."[19]

This long-standing situation has begun to improve during the past year. IO and PA have cooperated very effectively in dealing with various challenges—for example, the hijacked–tanker-truck bombing that caused civilian casualties, the fabricated charge that a female U.S. soldier threw a grenade at a crowd of civilians, and the repetition of that Taliban propaganda warhorse, the desecration of the Koran by U.S. troops.[20] All of these examples involved PSYOP and PA func-

---

[16] Richter, 2009, p. 108.

[17] DoD, 2003, p. 15.

[18] DoD, 2003, pp. 16, 27.

[19] Svet, 2010, p. 2.

[20] For example, in May 2008, two protesters were killed in Chaghcharan, the capital of Afghanistan's Ghor province, during a riot sparked by reports of USMIL desecration of a Koran ("Two Killed as Afghans Protest US Troop's Quran Desecration," DPA, May 22, 2008). In January 2010, the Taliban tried to foment a riot in the Gamser district of Helmand province over new allegations of USMIL desecrations of the Koran. Violence broke out, but

tions and were resolved favorably.[21] These successes suggest that that this type of coordination be institutionalized to be better prepared for timely counterpropaganda.

Failure to Exploit the Informal, Oral Afghan Communication Tradition

Information moves most efficiently in Afghanistan by word of mouth: gossip and face-to-face communication within a network of relatives, friends, and neighbors. According to one IO officer interviewed for this study, "information moves along human chains" in Afghanistan. Yet, in an IO planning session, "nobody ever says, 'How do we get a good piece of gossip into the system?'" One aspect of this issue to consider is whether a group of people can be identified who are mobile and who spread information using the traditional methods. In the judgment of some IO officers, large portions of the rural population are not receiving PSYOP messages, and the U.S. military has not figured out how to reach them.[22]

Survey data indicate that a considerable portion of Afghan respondents rely on word of mouth for information. In the Asia Foundation survey, respondents addressed the question, "If you wanted to find out about something important happening in your community, who, outside your family, would you want to tell you?" In Helmand, a plurality of 22 percent answered "a Village Chief/Community Leader," while 20 percent answered "a friend." In Kandahar, 35 percent of respondents answered "a friend," while 26 percent answered "neighbors/

---

it was quelled before it could attract mass participation (Richard Tomkins, "Anti-American Riot Rocks Afghan Town," *Human Events*, January 13, 2010).

[21] Interviews with two marines who were deployed to Gamser during that period indicated that U.S. forces reacted proactively and succeeded in defusing tensions by relying on the quick reaction of local community leaders and GIRoA officials with whom they had previously developed good relations.

[22] Interviews with IO personnel; author's personal observations in Afghanistan, 2009–2010.

villagers," and nationwide, 26 percent of respondents said "a friend," while 26 percent of respondents said "neighbors/villagers."[23]

Lack of Measures of Effectiveness

In his assessment of OEF, Major Cox concluded that current IO doctrine has provided little guidance on how to assess effectiveness, so assessments have usually been associated with battle-damage assessment. There has been no provision for assessing targets several days or even weeks after delivering a message. Finally, nonintelligence reporting that would help in obtaining an accurate assessment has not been readily available for analysis. Major Cox proposes the following steps to improve the situation:

- Develop accurate MOEs.
- Develop a collection plan that tracks the target audience and that can determine whether a delivery platform affected that audience.
- Develop measures of performance (MOPs) to assess the effectiveness of the delivery asset.

Because IO doctrine provides no guidance on developing MOEs, according to Major Cox, this process has been performed poorly. With little ability to measure the effectiveness of messages, Major Cox concludes that IO planners have received little or no feedback on whether PSYOP messages had the desired effect. Interviews with PSYOP advisors suggest that the unit intelligence cell could be a great resource to use in developing MOEs and developing a collection plan to determine how a particular platform affected the target audience. Closely related to the lack of MOEs is the lack of analysts who help devise and evaluate them. According to Major Cox, there have never been enough analysts to study the different reports generated by CA, PSYOP, and the other operators.[24] These reports should cover a wide gamut of informa-

---

[23] Rennie, Sharma, and Sen, 2008.

[24] Cox, 2006, p. 21.

tion, ranging from cultural sensitivities observed in specific tribal communities to effectiveness of different PSYOP initiatives. There have also not been enough trained analysts to handle the multiple reports and reporting formats that PSYOP can generate.[25]

Furthermore, because PSYOP messages have resided on one system and intelligence reports on the population have resided on another, analysts have not been able to consolidate all the information into one system. Finally, a lack of standardized databases has made it difficult for the Army or Marine Corps component intelligence staff officer (G-2) and IO section to share information. A standardized database, by utilizing the same terminology and search categories, would facilitate retrieving needed information more quickly.

A consensus exists among IO and PSYOP advisors interviewed for this study that, during the recent conflicts in Afghanistan and Iraq, Army doctrine lagged behind the tactics, techniques, and procedures that have become necessary to do the job. Due to the pace of operations, doctrine fell behind situations in the field. Doctrine should prepare commanders to integrate IO effectively into operations. Instead, in many cases, IO was treated as a stovepipe and only vertically integrated into staff planning. This is diametrically opposed to how the insurgents conduct their propaganda campaigns. The Taliban not only integrate their operations closely with IO but, in most cases, plan operations primarily for their IO effect on target audiences.

In his study, Major Cox documented the lack of IO integration with operations. According to him, integration of IO was usually left to the Fire Support Element, which usually did not have the training nor wider perspective to conduct IO successfully.[26] He identified two other trends worth noting:

- micro management by the command: Because of the probability of a single incident having international repercussions, commanders tend to overcontrol IO. Commanders did not just want to set the conditions for success with subordinates in IO; they wanted

---

[25] Cox, 2006, p. 21.

[26] Cox, 2006, p. 21.

to limit how the subordinates control IO. Unresponsive IO elements at HQ became irrelevant in the IO fight.

- commanders' view of IO as serving their political needs: Commanders view IO as the functions that will cover stories in the international media that meet the commanders' wants or desires, whereas current COIN doctrine emphasizes winning hearts and minds of the natives of contested areas. Complicating matters further is the fact that many combat commanders have difficulty grasping how the media operates and how to use the media to their advantage.[27]

It should be emphasized, however, that some IO officers point out that Cox's study was valid in 2006 when it was published but that major changes have been instituted since then. For example, commanders of the regional commands in Afghanistan enjoy great leeway in how to pursue IO, and there are significant variations among the commands. Others view Cox's critique as still relevant, despite the improvements that have occurred.

Some IO and PSYOP advisors interviewed for this study suggested that staffs prepare in IO preparation of the battlefield document, which should be updated constantly. The document should include information about the leadership and centers of gravity of the enemy, the government and local tribal and religious leadership, media platforms, and so forth. A clear picture of the IO environment would help the commander tremendously.

The 2003 DoD *Information Operations Roadmap* was more explicit in this regard, calling specifically for more-focused analytic and intelligence support. The road map noted,

> Combatant Command staffs lack organic capability to rapidly analyze complex systems and generate IO target sets. They need support from a robust analytical center that combines multidiscipline analysis capability with specifically tailored intelligence

---

[27] Cox, 2006, p. 55.

supporting IO. . . . Combatant Command staffs can not currently produce rapid solutions for tailored IO effects.[28]

This lack of adequate intelligence support, in turn, negatively affects the effort to develop MOEs.

---

[28] DoD, 2003, pp. 12, 38.

CHAPTER SEVEN

New Initiatives Being Implemented to Improve Psychological Operations

Two major initiatives to improve the efficacy of USMIL IO are under way that should be highlighted: the revision of IO doctrine and the announcement of a new multimedia strategy.

Revision of Information Operations Doctrine

The Army is presently making doctrinal changes to improve IO:

> Under President Barack Obama's directive, the army is rewriting its information operations manual. [LTC] Shawn Stroud, who until May 2009 served as director of strategic communication at U.S. Army Combined Arms Center in Fort Leavenworth, Kansas—which is coordinating the update—says previous versions of the army information doctrine gave senior officers far from the battlefield the responsibility for making decisions on communication and outreach. [The new manual will] "empower commanders" closer to the fight.

This is in keeping with the 2003 *Information Operations Roadmap* recommendation calling for greater authority for executing the IO mission to be given to the combatant commanders.[1] Afghanistan poses an especially pressing need for swifter communication decisions because

---

[1] DoD, 2003.

Taliban fighters—who often accuse U.S. troops of killing civilians during operations—are believed to stage civilian deaths and post videos of the fabricated footage. Stroud says U.S. field commanders need the tools to combat counterproductive messaging quickly, like speaking directly to the news media or even filming operations and posting their own combat footage online before the Taliban can. "It's almost like we've surrendered the information battlefield and said, 'Well, we don't play by the same rules as them because we have to tell the truth,' Stroud says. "The key is, we've got to be first with the truth. So we've got to build systems that do that."[2]

In his 2009 article "The Future of Information Operations," Major Richter made the following commentary on the effort to build new systems:

> The U.S. Army is revising Field Manual (FM) 3-13, *Information Operations*, further refining the November 2003 edition. Even so, its proposed doctrinal changes are evolutionary rather than revolutionary and frequently do not reflect commanders' operational experiences, appearing at times to address Cold War–era threat models
>
> Will the Army's new doctrinal definition and core capabilities of IO be adequate to support a national strategic communication plan? Will it be able to counter emergent and future threats?
>
> Unfortunately, the current definition and core capabilities of information operations appear inadequate to support a national strategic communications plan, counter emerging threats, or meet National Defense objectives over the next 15 years.
>
> Throughout U.S. agencies, including the military community, the concept of information operations in general and psychologi-

---

2 Bruno, 2009.

cal operations in particular as a weapon of deception has gradually diminished.[3]

Richter concludes his assessment with a series of recommendations for changes that merit attention but fall outside the scope of this study.

Regarding the ongoing rewrite of basic doctrine, in an interview for this monograph in 2009, LTC John "Chip" Bircher, former special assistant to the director of communication at ISAF, recommended amalgamating IO, PSYOP, and PA functions and producing them under a single communication-officer category. Thus, all personnel involved in this activity would have the same training and be part of the same organization, rather than being split, as is the current case. Although it clashes with IO orthodoxy, this monograph considers that it is a proposal worth considering seriously.

New Multimedia Strategy

To counter what has been an effective Taliban effort to stir up discontent in Afghanistan and Pakistan, President Obama supports the use of electronic media, cell phones, and radio to win the support of the Afghan populace. This approach seeks to take advantage of the enormous increase in cell-phone use in Afghanistan, including ubiquitous use among Taliban commanders.[4] RADM Gregory J. Smith, director of communication for U.S. Central Command, stated in an interview that success in this new endeavor will depend on the military's

> ability to deliver news quickly and accurately and equip locals with the tools to communicate freely with each other. Smith, who helped craft the Pentagon's definition of *strategic communication*, says an effective approach in Afghanistan could be "empowering

---

[3] Richter, 2009, p. 1.

[4] See "Satellite Backhaul Boosting Mobile Use in Afghanistan," Northern Sky Research, January 21, 2010.

conversation" among Afghans by supporting indigenous broadcasting, protecting radio towers, and fostering debate.[5]

In particular, Smith said, "possible new approaches include funding an expansion of radio transmission towers and news stations to allow local broadcasters to connect with indigenous publics. . . ."[6] Additional approaches include "protecting cell phone towers 'so more people can have access to cell phones to communicate amongst themselves through text messaging or just voice communications.'"[7] Cell-phone coverage is seen as another way to open up new lines of communication for people in remote, Taliban-dominated areas. The Taliban currently uses threats to force cell-phone providers to shut off service early at night, hampering local police and NGO efforts, cutting off the flow of information to locals, and preventing anyone from reporting insurgent movements or roadside bombs. Proposals to counter this trend include offering money or security to help protect privately owned cell phone towers, and perhaps even constructing cell phone towers on military bases. Given the Taliban concern over cell-phone towers and cell-phone use, this plan could give anti-Taliban forces an advantage:[8]

> Afghan officials say they support U.S. military efforts to improve communications capabilities as part of an overall effort to improve the GIRoA's image and counter the Taliban's messaging prowess. But that will not be easy, noted Michael Doran, a former deputy assistant secretary of defense, in a lecture on public diplomacy at the Heritage Foundation in February 2008. Doran said that in Afghanistan, U.S. forces carry out an operation "and within 26 minutes—we've timed it—the Taliban comes out with its version of what took place in the operation, which immediately finds its way on the tickers in the BBC at the bottom of the screen." The solution, Doran said, is much in line with what [Colonel] Stroud

---

[5] Bruno, 2009.

[6] Bruno, 2009.

[7] Bruno, 2009.

[8] Thom Shanker, "U.S. Plans a Mission Against Taliban's Propaganda," *New York Times*, August 15, 2009.

says the army is discussing—empowering U.S. and allied commanders to communicate more directly with local publics. . . . [T]he Pentagon is also considering jamming Taliban radio transmissions and disrupting militant websites, a strategy [Council on Foreign Relations] Senior Fellow Daniel Markey advocated in an August 2008 report and Pakistan's ambassador to the United States endorsed in an April 2009 *Wall Street Journal* op-ed. (The Afghan Taliban criticized the plan in a statement on its website.)[9]

Some experts suggest that, instead of blocking information, governments should disclose more and challenge Taliban motives and methods. In this respect, the Council on Foreign Relations' Stephen Biddle argues that coalition forces should consider focusing more on matching words with actions. "In places like Kunar Province, we have successfully designed integrated military-politico-economic operations to connect local Afghan populations with the government and create a political narrative that puts the Taliban on the outside, killing innocent Afghans, and ourselves on the inside, defending them," he says. Biddle says this strategy makes for "more effective communications" because words are matched by action.[10]

---

[9] Bruno, 2009. See also Michael Doran, remarks delivered at Public Diplomacy: Reinvigorating America's Strategic Communications Policy, a Heritage Foundation lecture on national security and defense, February 13, 2008, and Greg Bruno and Robert McMahon, "Afghan Defense Chief Unhappy with Obama Plan," Council on Foreign Relations, April 16, 2008.

[10] Bruno, 2009.

CHAPTER EIGHT

Recommendations for Improving the Effectiveness of Psychological Operations

Building on the recommendations made in the 2003 DoD *Information Operations Roadmap*, as well as other military and academic studies, this monograph recommends that the following actions be taken to improve PSYOP in Afghanistan and elsewhere.[1]

Hold a Lessons-Learned Conference of Information Operations and Psychological Operations Personnel

To assist in the ongoing DoD revision of IO and PSYOP doctrine and practice, a conference should be held attended primarily by IO and PSYOP personnel who have served in Afghanistan. The objective would be to define best practices based on their experiences and make recommendations for whatever reforms they believe should be made operationally, organizationally, and doctrinally. This would also include suggestions for training and personnel selection based on lessons learned.

Use Local Focus Groups to Pretest Messages

In assessing the effectiveness of PSYOP messages, the failure to take into account cultural, social, political, and religious factors was highlighted as one of the major deficiencies. As a partial remedy, the use of

---

[1] See DoD, 2003.

143

local focus groups is proposed to pretest messages, augmented public-opinion surveys for target-audience analysis, and use of key communicators to develop and disseminate messages. Pretesting is part of the standard PSYOP campaign planning cycle and should be rigorously implemented to avoid mistakes. Focus groups have been defined as simply a "structured conversation" and, in more detail, as "a carefully planned discussion designed to obtain perceptions on a defined area of interest in a permissive, nonthreatening environment."[2] During interviews conducted in January 2010 in Afghanistan, USMIL officers said that such focus groups have been set up in certain places. This is an excellent development. The overall recommendation is that this practice be standardized. Care must be taken, however, that the members of the focus group approximate the target audience as closely as possible. There have been cases in which the cultural-review function was given to Afghan government personnel or other individuals who were not from the area or even to Afghan Americans who have lived much of their lives in the United States.

PSYOP planners should also use focus groups as a pretest of a persuasive message or PSYOP product, providing feedback from a carefully selected sample of the target audience. Under ideal conditions, eight to 12 people should be selected to fit a targeted demographic or political cluster. This sharing of common characteristics creates a homogeneous group, which, in turn, fosters the permissive, nonthreatening environment needed for cohesion. A moderator

> adept at bringing out participants' responses through open-ended questions and associational techniques"[3] asks both general and specific questions. . . . With members seated around a table,

---

[2]   Richard A. Krueger, *Focus Groups: A Practical Guide for Applied Research*, Newbury Park, Calif.: Sage Publications, 1988, p. 18, as cited in Eric V. Larson, Richard E. Darilek, Daniel Gibran, Brian Nichiporuk, Amy Richardson, Lowell H. Schwartz, and Cathryn Quantic Thurston, *Foundations of Effective Influence Operations: A Framework for Enhancing Army Capabilities*, Santa Monica, Calif.: RAND Corporation, MG-654-A, 2009, p. 124.

[3]   Dennis W. Johnson, *No Place for Amateurs: How Political Consultants Are Reshaping American Democracy*, New York: Routledge, 2001, p. 102. (Footnote in original.)

the moderator works from a carefully prepared script, soliciting responses from participants in a variety of ways.[4]

In reality, in Afghanistan, it might not be possible to bring together eight to 12 people from the local community, but even two or three is better than none. The ideal procedure for running a focus group might also be impractical in the field. Nonetheless, it is useful to keep the technique in mind as PSYOP personnel implement a field-expedient version.

The basic idea is to get Afghan input into messages designed to influence Afghans. The focus group, to achieve its purpose, should be representative of the target audience. Therefore, if the target audience is villagers from a particular area of operations, then an effort should be made to get together a sample of that population sector. This can be done in the context of a CA event at which locals gather to receive assistance. Another option is to enlist the help of families of local soldiers or police officers. Obviously, participation must be voluntary. Sometimes, in conflictive areas, this is too hard to do. In such cases, local translators and members of their families could constitute the focus group. Care must be taken not to telegraph what response is wanted; otherwise, focus-group members will likely respond that all the leaflet designs or radio messages presented to them for evaluation are wonderful. The most-important part of this exercise is not only to get a reality check on prototypes of PSYOP products to be disseminated in particular areas but also to get input on changes in wording, or even ideas for entirely new products.

There are security issues here to be considered, which include the possibility that any message shown to a local focus group will leak before it is disseminated (which might or might not matter, depending on the particular message) and the safety of those involved. Bringing together a focus group needs to be done very discreetly, given the possibility of Taliban reprisals for collaborating with U.S. forces. For these reasons, the only practical pool of people to service in a focus group in a conflictive zone could be limited to the interpreters and ANSF

---

[4] Larson, Darilek, Gibran, et al., 2009, p. 124.

personnel who already associate openly with U.S. forces. They have their own prejudices and agendas, but they are Afghans, and having their input, however flawed it may be, is much better than producing PSYOP products with only Americans involved.

In his article on how to improve IO in Afghanistan, Commander LeGree emphasizes the need to avoid a U.S. perspective and advocates the use of the focus-group concept but in a more-informal manner:

> *It's in the delivery.* Clear examples of poor target audience analysis abound. The devil is certainly in the details, and these details can offend an audience if handled improperly. Adhere to the principles of immersion knowledge and local legitimacy. Bad information operations help the insurgents.
>
> *Our IO are often unsophisticated and clumsy.* As aforementioned, we frequently forget to listen to our audience and don't give them enough credit; or worse, we target the wrong audience. Remember, just because the people live simply does not mean they are simple. Focus information engagement strategies on that which the people care about and don't give unintended relevance to an enemy.
>
> *Seek a local opinion.* Do not disseminate IO or MISO products without a sanity check from Afghans from the area. Ask them questions, knowing that you will often get an answer of "what they think you want to hear." Wade through that and get a straightforward assessment.
>
> *Use a credible voice.* The best information operations come from respected Afghans with local credibility, not coalition forces. Quit falling in love with the guy who speaks English and deal with members of the community who command respect.[5]

---

[5] LeGree, 2010, p. 30.

Conduct Public-Opinion Surveys for Target-Audience Analysis and Posttesting

DoD recognizes the value of public-opinion surveys and is sponsoring a variety of efforts in this regard, including village-level polling. Significant work on human terrain mapping and cultural intelligence has also been conducted. However, much-better use of these data could be made to develop PSYOP themes and messages. The surveys should be keyed to specific PSYOP campaigns. Moreover, the emphasis should be on district-level polling (as opposed to national-level polls, which might not be representative of target audiences in conflictive areas). Survey research can provide quantitative baselines and trend analyses of key attitudes held by the target audience. In addition, they can help predict attitude change based on knowledge of underlying attitude structures and, thereby, help develop appropriately targeted messages.

Complementing focus groups, this type of survey can be used as a tool to develop persuasive messages. Survey research can provide quantitative baselines and trend analyses of key attitudes held by the target audience. In addition, surveys can help predict attitude change based on knowledge of underlying attitude structures and, thereby, help develop appropriately targeted messages. Larson and his colleagues add that surveys can clarify the relationship between subgroup characteristics and policy preferences, as well as the relative salience of policy issues (e.g., security, electric power, economic development).[6] Also, polling can be effective in posttesting specific PSYOP products, helping to determine if the audience reacts as intended.

Posttesting is part of standard PSYOP doctrine, and polling for that purpose should be part of every campaign plan, when feasible. Moreover, the Office of the Under Secretary of Defense for Policy recommends that assessment planning be given priority from the start. When this is done, products and messages have the opportunity to improve with feedback, adjustments, and time.[7]

---

[6] Larson, Darilek, Gibran, et al., 2009, p. 37.

[7] Friedly, 2010. On this point, see also Bebber, 2009.

Finally, it should be kept in mind that public-opinion surveys can be a psychological operation in themselves—that is, the questions themselves can be designed to evoke a desired emotion or thought. This is sometimes referred to as a *push poll*.[8] In such a situation, the actual answer given to the question is of secondary or no importance. An example of this type of survey question could be, "Do you believe allegations that Karzai himself ordered officials to stuff ballot boxes in Kandahar?" Even if the respondent answers negatively, being asked the question in the context of a survey gives credibility to the allegation and could induce doubts and suspicions.

It should be noted that polling and interviews do have pitfalls, particularly in conflictive areas, such as Afghanistan. A previous RAND study cautions,

> direct observation, polling, surveys, interviews, and other methods can be used to gauge the effectiveness of the shaping campaign.[9] Yet challenges remain. These techniques are difficult to get right and are expensive to implement. Additionally, they are subject to various forms of bias—including response bias (i.e., when the respondent tells the interviewer what he or she wants to hear), selection bias (i.e., when the sample is not chosen in a representative fashion), and self-selection bias (i.e., when only people who want to participate in a poll do so, and the responses of these individuals differ substantially from the hypothetical responses of those who did not participate).[10]

Utilize Key Communicators to Help Develop and Disseminate Messages

Using key communicators to disseminate messages is part of standard PSYOP doctrine as described in the PSYOP manual. The assumption is

---

[8]   See "What Is a 'Push' Poll?" American Association for Public Opinion Research, undated web page.

[9]   Lamb, 2005, p. 29. (Footnote in original.)

[10]   Helmus, Paul, and Glenn, 2007, pp. 47–48.

that messages are more credible if they come from a figure who already enjoys prestige within the target audience and is already considered a credible source of advice and information. In Afghanistan, key communicators can vary greatly between communities. They could be a mullah or *maulawi* (Islamic cleric), a traditional khan or malik (chief), an educated schoolteacher, a wealthy merchant known for providing charity, a local leader who maintains a loyal following, or a government official, among others. None of these people might be automatically disposed to work with the U.S. military, however, and it might be necessary to spend time convincing them that it is in their own interest, and the interest of their community, to cooperate.

The traditional PSYOP role of the key communicator should be expanded. Key communicators should be considered partners in developing messages, contributing not only to the wording but also to the content. As with everyone else in Afghanistan, these key communicators have their own agendas, so the PSYOP officer needs to become aware of their tribal, subtribal, political, or *qawm* (local social and economic grouping) loyalties. As with other suggestions, the key issue here is the need to integrate into the Afghan landscape as much as possible and use local collaborators to get messages and themes to target audiences.[11]

This proposed innovation is beginning to take place. There is a strong emphasis in ISAF currently on meetings with traditional leaders and interaction with tribal and local jirgas in which spontaneous opinions are expressed, both in favor of and against U.S. forces. This type of dialogue is an excellent first step in understanding the target audience and, ultimately, crafting better messages to influence that audience. In addition, a native Pashtun analyst who has helped the U.S. military implement PSYOP campaigns in Afghanistan argues that, besides traditional tribal leaders, other influential people in the community should be enlisted as key communicators, ranging from poets to schoolteachers.

---

[11] Author's personal observations in Afghanistan; FM 33-1, 1993; FM 33-1-1, 1994.

Harmonize Information-Operations Doctrine and Practice, and Implement Greater Integration with Psychological Operations and Public Affairs

The current disconnect between official IO doctrine and how IO are practiced in the field is counterproductive. The situation has been further complicated by the elimination of the term *PSYOP*, entailing, in the words of U.S. Special Operations Command (USSOCOM) commander ADM Eric T. Olson, a "complete change in organization, practice, and doctrine."[12] That being the case, at the time of this writing, clarification of the revised PSYOP mission is needed. Also, the current division between PSYOP and PA works to the advantage of Taliban propagandists, who routinely accuse U.S. forces of needlessly causing civilian casualties. Closer coordination between PSYOP and PA would enhance counterpropaganda effectiveness. As mentioned previously, the recommendation of an IO officer who served in ISAF to combine IO, PSYOP, and PA into a new military occupational specialty (MOS) of communication officer should be considered, so that everyone receives the same basic training and doctrine, enhancing operational unity.

---

[12] Paddock, 2010.

Plan for Campaign Against Improvised Explosive Devices

This appendix contains a good example of a PSYOP campaign plan against IEDs that seems to be effective in influencing target audiences.

Figure A.1
Information-Operations Support to the Counter–Improvised Explosive Device Fight

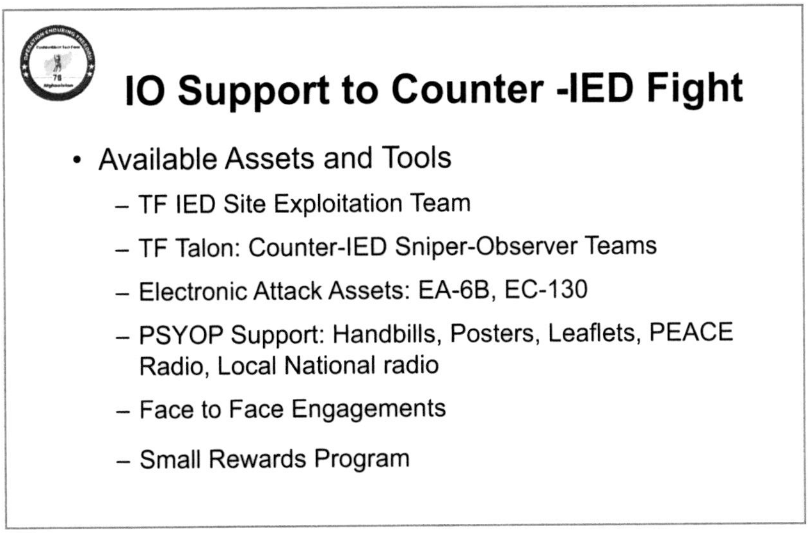

IO Support to Counter -IED Fight

- Available Assets and Tools
  - TF IED Site Exploitation Team
  - TF Talon: Counter-IED Sniper-Observer Teams
  - Electronic Attack Assets: EA-6B, EC-130
  - PSYOP Support: Handbills, Posters, Leaflets, PEACE Radio, Local National radio
  - Face to Face Engagements
  - Small Rewards Program

NOTE: TF = task force. EA-6B = Northrop Grumman Prowler, an electronic warfare aircraft. EC-130 = Lockheed cargo aircraft.
RAND *MG1060-A.1*

Figure A.2
Information-Operations Input to the Targeting Process

IO Input to Targeting Process

- Target Audience Analysis: *Who can you reach?*
  - ACM
  - Children
  - Mothers
  - Fathers
- Key Communicators/Spheres of Influence: *Who is the real decision maker?*
  - Provincial: Governor, Advisors, Mullahs, Police Chief, ANA
  - District: Shuras, Teachers, Doctors, Mullahs, Clans, ANA
  - Village: Elders, Clans, Children, Mullah, Teacher
- Tailored messaging for Target Audience: *What effect do you want, and how do you get it?*
  - Asset allocation
- Delivery mechanism: *Technical vs. Human*
  - *Soldiers, Leaders, Afghans*
  - *EA-6B, EC-130*
  - *PSYOP Radio, Print dissemination*
  - *Other assets*
- Measures of Effectiveness vs. Impact Indicators

RAND *MG1060-A.2*

Figure A.3
Information-Operations Themes and Messages

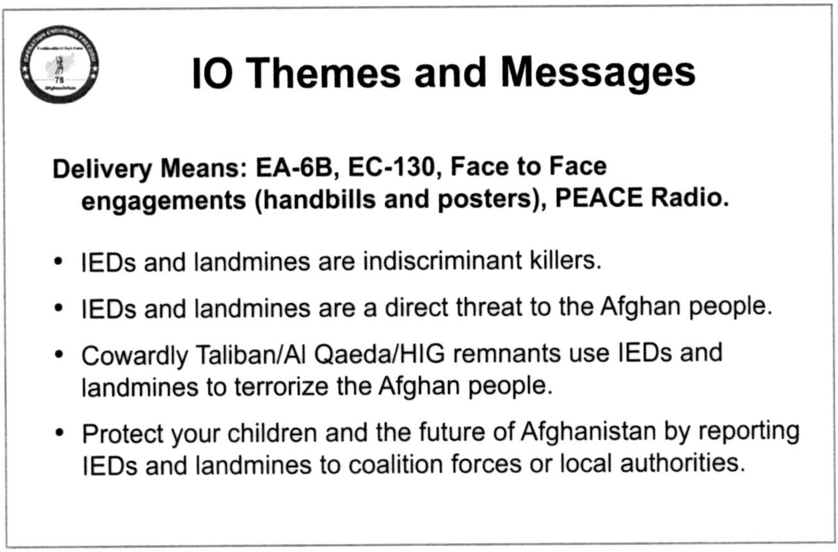

IO Themes and Messages

**Delivery Means: EA-6B, EC-130, Face to Face
engagements (handbills and posters), PEACE Radio.**

- IEDs and landmines are indiscriminant killers.

- IEDs and landmines are a direct threat to the Afghan people.

- Cowardly Taliban/Al Qaeda/HIG remnants use IEDs and
  landmines to terrorize the Afghan people.

- Protect your children and the future of Afghanistan by reporting
  IEDs and landmines to coalition forces or local authorities.

RAND *MG1060-A.3*

Figure A.4
Sample Products

Figure A.5
Impact Indicators

Impact Indicators

Impact Indicators are key events that, when analyzed together, provide subjective Measures of Effectiveness.

- Information provided by Afghan nationals on location of IEDs, IED cells, IED materiel
  - Small Rewards Program?
  - Goodwill?
  - Sense of Nationalism?
- RCIED Flash observances –EA6B, EC-130, local press reporting, local national feedback
- TF Talon (counter-IED snipers) employment / results
- Detainee interrogation feedback
- IED Site exploitation results
  - Posters at site
  - LN interviews
  - Small Rewards

NOTE: RCIED = radio-controlled improvised explosive device.
RAND *MG1060-A.5*

Campaign Plan to Support the 2004 Afghan Presidential Elections

Figure B.1
Election Exploitation Plan

SOURCE: Operation Enduring Freedom Combined Joint Task Force 76–
Afghanistan.
RAND *MG1060-B.1*

Figure B.2
Shaping the Environment

 # Shaping the Environment

Effects Desired:

- ACM is discredited and ACM actions are seen as a desperate attempt to jeopardize first the elections and ultimately Afghanistan's future.
- UNAMA perception of security is improving and sufficient for successful elections.
- International media highlights the successes of the Coalition efforts to enable Afghan institutions.

NOTE: UNAMA = United Nations Assistance Mission in Afghanistan.
RAND *MG1060-B.2*

Figure B.3
Whom the United States Needs to Influence

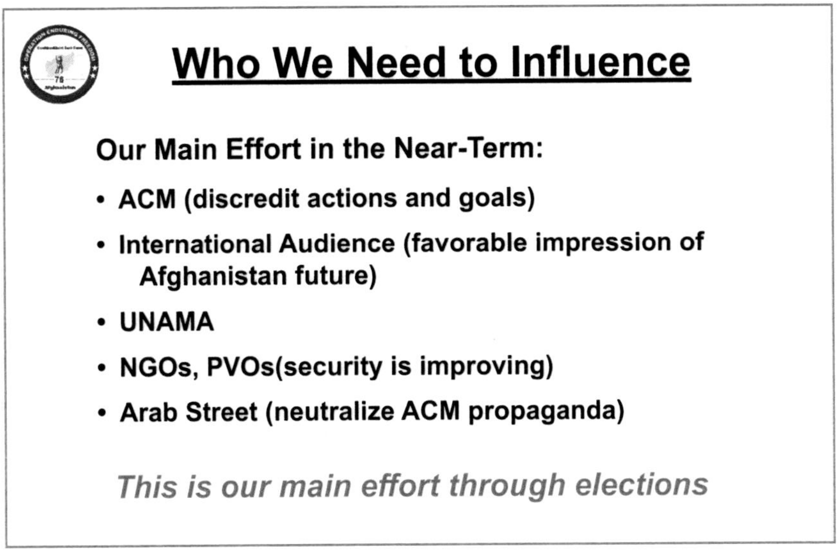

<u>Who We Need to Influence</u>

Our Main Effort in the Near-Term:

- ACM (discredit actions and goals)
- International Audience (favorable impression of Afghanistan future)
- UNAMA
- NGOs, PVOs(security is improving)
- Arab Street (neutralize ACM propaganda)

This is our main effort through elections

NOTE: PVO = private voluntary organization.
RAND MG1060-B.3

Figure B.4
Target Audiences for Election Success

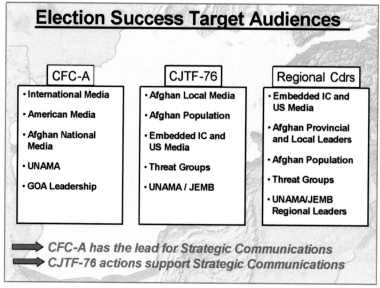

Figure B.5
Election Exploitation Themes

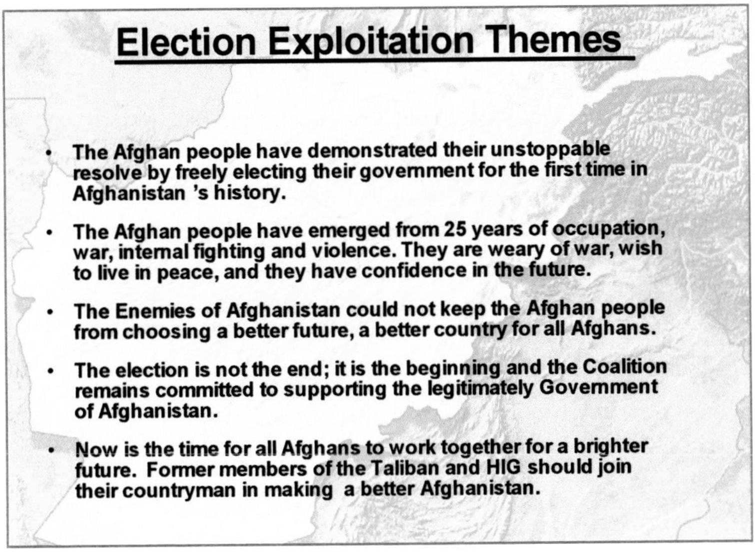

RAND *MG1060-B.5*

Figure B.6
Election Information Plan

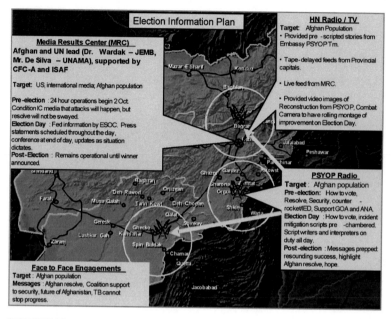

RAND MG1060-B.6

Figure B.7
Postelection Exploitation

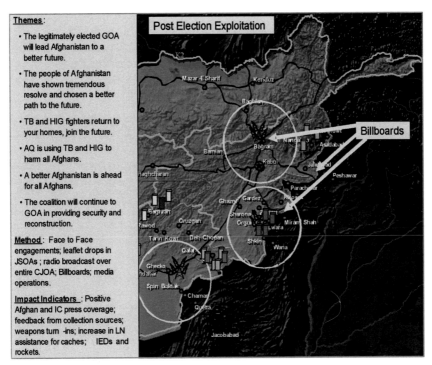

NOTE: AQ = al-Qa'ida. JSOA = joint special operations area.

RAND *MG1060-B.7*

Bibliography

1st Battalion, 5th Marines, "COIN in Helmand Province: After the Clear—Thoughts and Tips on Non Kinetic Actions," undated briefing. As of December 23, 2010:
http://usacac.army.mil/cac2/coin/repository/COIN_Helmand_Province_AAR_via_photos_23_Oct_(NXPowerLite).pptx

Abdul-Ahad, Ghaith, "New Evidence of Widespread Fraud in Afghanistan Election Uncovered," *Guardian*, September 19, 2009. As of January 18, 2011:
http://www.guardian.co.uk/world/2009/sep/18/afghanistan-election-fraud-evidence

Afghan Center for Socio-Economic and Opinion Research, poll for ABC News, BBC, and ARD, December 11–23, 2009. As of January 18, 2011:
http://news.bbc.co.uk/2/shared/bsp/hi/pdfs/11_01_10_afghanpoll.pdf

"Afghan Support for Karzai's Government Low: Pentagon Report," Agence France-Presse, April 29, 2010.

"Afghan Teacher Shot Dead After Condemning Suicide Bombings as Un-Islamic," *Daily Mail*, May 14, 2008. As of January 18, 2011:
http://www.dailymail.co.uk/news/article-566439/Afghan-teacher-shot-dead-condemning-suicide-bombings-Islamic.html

"Afghanistan: National Opinion Poll for BBC, ABC News and ARD," press release, BBC Press Office, February 9, 2009.

Aronson, Elliot, Timothy D. Wilson, and Robin M. Akert, *Social Psychology*, 3rd ed., New York: Longman, 1999.

Azami, Dawood, "Taliban Slick Propaganda Confronts US," *BBC News*, August 3, 2009. As of December 22, 2010:
http://news.bbc.co.uk/2/hi/south_asia/8176259.stm

Bagram Media Center, "Regional Command East Boosts Security, with Afghan Participation in DoD's Reward Program," press release, April 4, 2010. As of December 28, 2010:
http://www.cjtf82.com/en/press-releases-mainmenu-326/2614-regional-command-east-boosts-security-with-afghan-participation-in-dods-reward-program.html

Baker, COL Ralph O., "The Decisive Weapon," *Military Review*, May–June 2006, pp. 13–32.

Bebber, Ensign Robert J., "Developing an IO Environmental Assessment in Khost Province: Information Operations at PRT Khost in 2008," *Small Wars Journal*, February 28, 2009. As of December 22, 2010:
http://smallwarsjournal.com/blog/2009/02/developing-an-io-environmental/

Bokhari, Farhan, "U.S. Envoy: Taliban Can't Stop Afghan Elections," *CBS News*, August 17, 2009. As of December 22, 2010:
http://www.cbsnews.com/8301-503543_162-5247069-503543.html

———, "After Attack, Pakistan Confronts Challenge of Burqa-Clad Bombers," *World Watch*, December 26, 2010. As of December 28, 2010:
http://www.cbsnews.com/
8301-503543_162-20026610-503543.html?tag=contentMain;contentBody

Bruno, Greg, *Winning the Information War in Afghanistan and Pakistan*, New York: Council on Foreign Relations, May 11, 2009. As of December 22, 2010:
http://www.cfr.org/publication/19330/
winning_the_information_war_in_afghanistan_and_pakistan.html

Bruno, Greg, and Robert McMahon, "Afghan Defense Chief Unhappy with Obama Plan," Council on Foreign Relations, April 16, 2008. As of December 22, 2010:
http://www.cfr.org/publication/19116/
afghan_defense_chief_unhappy_with_obama_plan.html

Bumiller, Elisabeth, and Carlotta Gall, "U.S. Admits Civilians Died in Afghan Raids," *New York Times*, May 7, 2009. As of December 29, 2010:
http://www.nytimes.com/2009/05/08/world/asia/08afghan.html

Burns, CDR Ed, U.S. Navy, Joint Information Operations Center, interview with Christopher Paul and Todd C. Helmus, Lackland AFB, Tex., February 16, 2006.

Campbell, Donald Thomas, and Julian C. Stanley, *Experimental and Quasi-Experimental Designs for Research*, Boston, Mass.: Houghton Mifflin, 1963.

"Commander of Army Civil Affairs and Psychological Operations Command Leaving for New Job," Associated Press, August 20, 2009.

"Commando Solo Radio Scripts: War on Terrorism in Afghanistan," undated.

Cox, MAJ Joseph L., U.S. Army, *Information Operations in Operations Enduring Freedom and Iraqi Freedom: What Went Wrong?* Fort Leavenworth, Kan.: School of Advanced Military Studies, U.S. Army Command and Staff College, 2006. As of December 22, 2010:
http://www.fas.org/irp/eprint/cox.pdf

Davis, CPT Richard, "CA/PSYOPS [sic] in Afghanistan," *Infantry Online*, April 15, 2003.

"Deoband Ulema Term All Taliban Actions Un-Islamic," *Dawn*, June 20, 2009.

"Despite Deep Challenges in Daily Life, Afghans Express a Positive Outlook," *Life in Afghanistan*, ABC News poll, December 7, 2005. As of December 22, 2010:
http://www.charneyresearch.com/pdf/998a1Afghanistan.pdf

DoD—*See* U.S. Department of Defense.

Doran, Michael, remarks delivered at Public Diplomacy: Reinvigorating America's Strategic Communications Policy, a Heritage Foundation lecture on national security and defense, February 13, 2008. As of December 22, 2010:
http://www.heritage.org/research/lecture/
public-diplomacy-reinvigorating-americas-strategic-communications-policy

Dorronsoro, Gilles, *The Taliban's Winning Strategy in Afghanistan*, Washington, D.C.: Carnegie Endowment for International Peace, 2009. As of December 22, 2010:
http://www.carnegieendowment.org/files/taliban_winning_strategy.pdf

Eassa, LTC Charles, Deputy Director, U.S. Army Information Operations, as quoted in Michael Schrage, "Use Every Article in the Arsenal: Good Press Is a Legitimate Weapon," *Washington Post*, January 15, 2006. As of December 28, 2010:
http://www.washingtonpost.com/wp-dyn/content/article/2006/01/13/
AR2006011302303.html

Eckel, Mike, "U.S. Military Turns to Video of 9/11/01 to Win Hearts and Minds of Afghans," Associated Press, 2002.

Economist Intelligence Unit, *Afghanistan: Country Profile*, London, 2008.

Elson, Sara Beth, *Effectiveness of U.S. Information Operations in Afghanistan*, unpublished research, Santa Monica, Calif.: RAND Corporation, 2009.

"Excerpts from Afghan President Hamid Karzai's Interview with *The Washington Post*," *Washington Post*, November 14, 2010. As of January 18, 2011:
http://www.washingtonpost.com/wp-dyn/content/article/2010/11/14/
AR2010111400002.html?sid=ST2010111305091

"Exploding Misconceptions: Alleviating Poverty May Not Reduce Terrorism but Could Make It Less Effective," *Economist*, December 16, 2010. As of January 18, 2011:
http://www.economist.com/node/17730424

FM 3-05.30—*See* Headquarters, Department of the Army, 2005.

FM 33-1—*See* U.S. Marine Corps and U.S. Department of the Army, 1993.

FM 33-1-1—*See* Headquarters, Department of the Army, 1994.

Friedly, Douglas, senior analyst, Office of the Under Secretary of Defense for Policy, Information Operations Directorate, comments on an earlier draft of this monograph, 2010.

Friedman, SGM (ret.) Herbert A., "Psychological Operations in Afghanistan," *psywarrior.com*, undated web page (a). As of December 22, 2010:
http://www.psywarrior.com/Herbafghan.html

———, "PSYOP Dissemination," *psywarrior.com*, undated web page (b). As of December 22, 2010:
http://www.psywarrior.com/dissemination.html

———, "The Strange Case of the Vietnamese 'Late Hero' Nguyen Van Be," *psywarrior.com*, undated web page (c). As of December 22, 2010:
http://www.psywarrior.com/BeNguyen.html

Galbraith, Peter W., "What I Saw at the Afghan Election," *Washington Post*, October 4, 2009. As of January 18, 2011:
http://www.washingtonpost.com/wp-dyn/content/article/2009/10/02/AR2009100202855.html

Gannon, Kathy, "Afghans Blame Both US, Taliban for Insecurity," Associated Press, April 16, 2010.

Gant, MAJ Jim, U.S. Army, *One Tribe at a Time*, Los Angeles, Calif.: Nine Sisters Imports, 2009.

Garamone, Jim, "U.S. Commando Solo II Takes Over Afghan Airwaves," American Forces Press Service, October 29, 2001.

Headquarters, Department of the Army, *Psychological Operations Techniques and Procedures*, Washington, D.C., Field Manual 33-1-1, May 5, 1994.

———, *Information Operations: Doctrine, Tactics, Techniques, and Procedures*, Field Manual 3-13, November 2003.

———, *Psychological Operations*, Field Manual 3-05.30, Marine Corps Reference Publication 3-40.6, April 2005. As of December 22, 2010:
http://www.fas.org/irp/doddir/army/fm3-05-30.pdf

Helmus, Todd C., Christopher Paul, and Russell W. Glenn, *Enlisting Madison Avenue: The Marketing Approach to Earning Popular Support in Theaters of Operation*, Santa Monica, Calif.: RAND Corporation, MG-607-JFCOM, 2007. As of December 22, 2010:
http://www.rand.org/pubs/monographs/MG607.html

Hoffman, Bruce, "Today's Highly Educated Terrorists," *National Interest*, September 15, 2010. As of January 18, 2011:
http://nationalinterest.org/blog/bruce-hoffman/
todays-highly-educated-terrorists-4080

Holbrooke, Richard, press briefing on the new strategy for Afghanistan and Pakistan, Washington, D.C., March 27, 2009. As of December 23, 2010:
http://www.whitehouse.gov/the_press_office/Press-Briefing-by-Bruce-Riedel-Ambassador-Richard-Holbrooke-and-Michelle-Flournoy-on-the-New-Strategy-for-Afghanistan-and-Pakistan/

ICOS—*See* International Council on Security and Development.

International Council on Security and Development, *Afghanistan: The Relationship Gap*, Brussels, July 2010. As of December 28, 2010:
http://www.icosgroup.net/modules/reports/afghanistan_relationship_gap

"Is General McChrystal a Hippie?" *Economist*, August 27, 2009. As of December 23, 2010:
http://www.economist.com/blogs/democracyinamerica/2009/08/
is_general_mcchrystal_a_hippie

"ISAF to Continue Night Raids in Afghanistan," *TOLOnews.com*, November 29, 2010. As of January 18, 2011:
http://www.tolonews.com/en/
afghanistan/1181-isaf-to-continue-night-raids-in-afghanistan-

Johnson, Dennis W., *No Place for Amateurs: How Political Consultants Are Reshaping American Democracy*, New York: Routledge, 2001.

"Joint Group Brings Aid to Villages in Helmand," press release, International Security Assistance Force Afghanistan, December 26, 2010. As of January 18, 2011:
http://isaf-live.webdrivenhq.com/article/isaf-releases/
joint-group-brings-aid-to-villages-in-helmand-2.html

Jones, M., *Intelligence and Counterinsurgency: The Malayan Experience*, 2009.

Josten, Richard J., "Strategic Communication: Key Enabler for Elements of National Power," *IO Sphere*, Summer 2006, pp. 16–20. As of December 23, 2010:
http://www.carlisle.army.mil/DIME/documents/iosphere_summer06_josten.pdf

JP 1-02—*See* U.S. Joint Chiefs of Staff, 2001 (2010).

JP 3-53—*See* U.S. Joint Chiefs of Staff, 2003.

Kalton, Graham, *Introduction to Survey Sampling*, Beverly Hills, Calif.: Sage Publications, 1983.

Kelly, Jon, "The Secret World of 'Psy-Ops,'" *BBC News*, June 20, 2008. As of December 22, 2010:
http://news.bbc.co.uk/2/hi/uk_news/7464430.stm

Knowlton, Brian, and Judy Dempsey, "U.S. Adviser Holds Firm on Airstrikes in Afghanistan," *New York Times*, May 10, 2009. As of December 29, 2010:
http://www.nytimes.com/2009/05/11/world/asia/11karzai.html

Koontz, Christopher N., *Enduring Voices: Oral Histories of the U.S. Army Experience in Afghanistan, 2003–2005*, Washington, D.C.: Center of Military History, U.S. Army, 2008.

Krueger, Alan B., and Jitka Malecková, "Education, Poverty and Terrorism: Is There a Causal Connection?" *Journal of Economic Perspectives*, Vol. 17, No. 4, Fall 2003, pp. 119–144.

Krueger, Richard A., *Focus Groups: A Practical Guide for Applied Research*, Newbury Park, Calif.: Sage Publications, 1988.

Lamb, Christopher J., *Review of Psychological Operations Lessons Learned from Recent Operational Experience*, Washington, D.C.: National Defense University Press, September 2005. As of December 23, 2010:
http://purl.access.gpo.gov/GPO/LPS82542

Larson, Eric V., Richard E. Darilek, Daniel Gibran, Brian Nichiporuk, Amy Richardson, Lowell H. Schwartz, and Cathryn Quantic Thurston, *Foundations of Effective Influence Operations: A Framework for Enhancing Army Capabilities*, Santa Monica, Calif.: RAND Corporation, MG-654-A, 2009. As of December 22, 2010:
http://www.rand.org/pubs/monographs/MG654.html

Larson, Eric V., Richard E. Darilek, Dalia Dassa Kaye, Forrest E. Morgan, Brian Nichiporuk, Diana Dunham-Scott, Cathryn Quantic Thurston, and Kristin J. Leuschner, *Understanding Commanders' Information Needs for Influence Operations*, Santa Monica, Calif.: RAND Corporation, MG-656-A, 2009. As of December 23, 2010:
http://www.rand.org/pubs/monographs/MG656.html

Larawbar, untitled video, January 12, 2011. As of January 18, 2011:
http://www.youtube.com/watch?v=-rLzOGTGtTY

LeGree, CDR Larry, U.S. Navy, "Thoughts on the Battle for the Minds: IO and COIN in the Pashtun Belt," *Military Review*, September–October 2010, pp. 21–32. As of December 28, 2010:
http://usacac.army.mil/CAC2/MilitaryReview/Archives/English/
MilitaryReview_20101031_art006.pdf

Lemos, Charles, "The Mullen Doctrine," *My Direct Democracy*, March 14, 2010. As of December 23, 2010:
http://mydd.com/2010/3/14/the-mullen-doctrine

Main, COL Francis Scott, U.S. Army Reserve, *Psychological Operations Support to Strategic Communications in Afghanistan*, Carlisle Barracks, Pa.: U.S. Army War College, strategy research project, March 24, 2009. As of December 23, 2010:
http://handle.dtic.mil/100.2/ADA497711

McChrystal, GEN Stanley A., *ISAF Commander's Counterinsurgency Guidance*, Kabul: Headquarters, International Security Assistance Force, August 2009. As of December 22, 2010:
http://www.nato.int/isaf/docu/official_texts/counterinsurgency_guidance.pdf

McGivering, Jill, "Afghan People 'Losing Confidence,'" *BBC News*, February 9, 2009. As of January 18, 2011:
http://news.bbc.co.uk/2/hi/south_asia/7872353.stm

Meyerle, Jerry, Megan Katt, and Jim Gavrilies, *Counterinsurgency on the Ground in Afghanistan: How Different Units Adapted to Local Conditions*, Center for Naval Analyses, November 2010.

Mullen, ADM Michael G., chair, U.S. Joint Chiefs of Staff, "Strategic Communication: Getting Back to Basics," *Joint Forces Quarterly*, Vol. 55, 4th Quarter, August 28, 2009. As of December 22, 2010:
http://www.jcs.mil/newsarticle.aspx?ID=142

Myers, Gen Richard B., Chairman of the Joint Chiefs of Staff, "Policy on Public Affairs Relationship to Information Operations," memorandum, CM-2077-04, September 24, 2004.

Mynott, Adam, "Afghans More Optimistic for Future, Survey Shows," *BBC News*, January 11, 2010. As of December 28, 2010:
http://news.bbc.co.uk/2/hi/8448930.stm

Ngo, Dong, "U.S. Military Joins Twitter, Facebook," *CNET News*, June 1, 2009. As of January 18, 2011:
http://news.cnet.com/8301-17939_109-10253555-2.html

Operation Enduring Freedom, Combined Joint Task Force 76–Afghanistan, "Information Operations: Command Themes Week of 19–25 January 2005," briefing, 2005.

Paddock, Alfred Jr., "PSYOP: On a Complete Change in Organization, Practice, and Doctrine," *Small Wars Journal*, June 26, 2010. As of December 22, 2010:
http://smallwarsjournal.com/blog/journal/docs-temp/463-paddock.pdf

Partlow, Joshua, and Scott Wilson, "Karzai Rails Against Foreign Presence, Accuses West of Engineering Voter Fraud," *Washington Post*, April 2, 2010. As of January 18, 2011:
http://www.washingtonpost.com/wp-dyn/content/article/2010/04/01/AR2010040101681.html

Paul, Christopher, *Information Operations: Doctrine and Practice—A Reference Handbook*, Westport, Conn.: Praeger Security International, 2008.

———, *Whither Strategic Communication? A Survey of Current Proposals and Recommendations*, Santa Monica, Calif.: RAND Corporation, OP-250-RC, 2009. As of December 23, 2010:
http://www.rand.org/pubs/occasional_papers/OP250.html

————, "'Strategic Communication' Is Vague: Say What You Mean," *Joint Force Quarterly*, Vol. 56, 1st Quarter, 2010, pp. 10–13. As of December 23, 2010: http://www.au.af.mil/au/awc/awcgate/jfq/paul_sc_is_vague.pdf

Qaddafi, Muammar, *The Green Book*, Ottawa: Jerusalem International Publishing on behalf of the Green Book World Center for Research and Study, Tripoli, Socialist People's Libyan Arab Jamahiriya, 1983.

Rennie, Ruth, Sudhindra Sharma, and Pawan Kumar Sen, *Afghanistan in 2008: A Survey of the Afghan People*, Kabul: Asia Foundation, Afghanistan Office, 2008.

Richter, MAJ Walter E., U.S. Army, "The Future of Information Operations," *Military Review*, January–February 2009, pp. 103–113. As of December 23, 2010: http://usacac.army.mil/CAC2/MilitaryReview/Archives/English/ MilitaryReview_20090228_art013.pdf

Roberts, M. E., *Villages of the Moon: Psychological Operations in Southern Afghanistan*, Baltimore, Md.: Publish America, 2005.

Ruttig, Thomas, *The Other Side: Dimensions of the Afghan Insurgency—Causes, Actors, an[d] Approaches to "Talks,"* Afghanistan Analysts Network, July 2009. As of December 23, 2010: http://aan-afghanistan.com/uploads/200907%20AAN%20Report%20Ruttig%20 -%20The%20Other%20Side.PDF

Saber, Shapoor, Fetrat Zerak, and Abaceen Nasimi, "Soul-Searching Following Farah Tragedy," Institute for War and Peace Reporting, June 5, 2009.

"Satellite Backhaul Boosting Mobile Use in Afghanistan," Northern Sky Research, January 21, 2010. As of January 18, 2011: http://www.nsr.com/index.php?option=com_content&view=article&id=308&catid =94&Itemid=153

Schanz, Marc V., "The New Way of Psyops," *Air Force Magazine*, Vol. 93, No. 11, November 2010. As of January 18, 2011: http://www.airforce-magazine.com/MagazineArchive/Pages/2010/ November%202010/1110psyops.aspx

Schlussel, Debbie, "Weekend Read: Study Shows Terrorists Are More Educated, Wealthy, from Oppressive Countries," *Debbie Schlussel*, July 6, 2007. As of January 18, 2011: http://www.debbieschlussel.com/1521/weekend-read-study-shows-terrorists-are- more-educated-wealthy-from-oppressive-countries/

Scott, Colonel Jeffrey, *Speed Versus Accuracy: A Zero Sum Game*, unpublished paper, 2010.

Shanker, Thom, "U.S. Plans a Mission Against Taliban's Propaganda," *New York Times*, August 15, 2009. As of December 22, 2010: http://www.nytimes.com/2009/08/16/world/asia/16policy.html

Sholtis, Maj Tadd, "Planning for Legitimacy: A Joint Operational Approach to Public Affairs," *Air and Space Power Journal*, June 8, 2005. As of December 29, 2010:
http://www.airpower.maxwell.af.mil/airchronicles/cc/sholtis.html

Singer, Peter W., *Winning the War of Words: Information Warfare in Afghanistan*, Washington, D.C.: Brookings Institution, Analysis Paper 5, October 23, 2001. As of December 22, 2010:
http://www.brookings.edu/papers/2001/1023afghanistan_singer.aspx

Smith, RADM Gregory J., "Countering the Taliban's Message in Afghanistan and Pakistan," interview with Greg Bruno, staff writer, Council on Foreign Relations, May 11, 2009. As of December 29, 2010:
http://www.cfr.org/publication/19257/
countering_the_talibans_message_in_afghanistan_and_pakistan.html

Spiegel, Peter, Jonathan Weisman, and Yochi J. Dreazen, "Obama Bets Big on Troop Surge," *Wall Street Journal*, December 2, 2009. As of January 18, 2011:
http://online.wsj.com/article/SB125967363641871171.html

Strategic Communication Laboratories, "Perceptions of the Afghan National Police (ANP) in Arghandab and Maywand Districts, Kandahar Province, Afghanistan," London, undated.

Sudman, Seymour, *Applied Sampling*, New York: Academic Press, 1976.

"Support for U.S. Efforts Plummets Amid Afghanistan's Ongoing Strife," *Afghanistan: Where Things Stand*, ABC News/BBC/ARD poll, February 9, 2009. As of December 22, 2010:
http://abcnews.go.com/images/PollingUnit/1083a1Afghanistan2009.pdf

Svet, Oleg, "Fighting for a Narrative: A Campaign Assessment of the US-Led Coalition's Psychological and Information Operations in Afghanistan," *Small Wars Journal*, September 12, 2010. As of December 29, 2010:
http://smallwarsjournal.com/blog/2010/09/fighting-for-a-narrative/

"Taliban Blamed for Acid Attack on Afghan Schoolgirls," Associated Press, November 14, 2008.

Tomkins, Richard, "Anti-American Riot Rocks Afghan Town," *Human Events*, January 13, 2010. As of January 18, 2011:
http://www.humanevents.com/article.php?id=35171

Tran, Mark, "US Military Admits Errors in Air Strikes That Killed Scores of Afghan Civilians," *Guardian*, June 3, 2009. As of December 29, 2010:
http://www.guardian.co.uk/world/2009/jun/03/afghanistan-us-airstrikes-errors

Turner, COL Kenneth A., U.S. Army, Commanding Officer, 4th Psychological Operations Group, interview with Todd C. Helmus and Christopher Paul, Ft. Bragg, N.C., December 14, 2005.

Tversky, Amos, and Daniel Kahneman, "The Framing of Decisions and the Psychology of Choice," *Science*, Vol. 211, No. 4481, 1981, pp. 453–458.

"Two Killed as Afghans Protest US Troop's Quran Desecration," DPA, May 22, 2008.

"U.N. Official Admits Afghan Vote Fraud," *CNN World*, October 11, 2009. As of January 18, 2011:
http://articles.cnn.com/2009-10-11/world/afghanistan.
un_1_kai-eide-karzai-allegations-of-election-fraud?_s=PM:WORLD

U.S. Department of the Army, *Operations*, Washington, D.C.: Headquarters, Department of the Army, Field Manual 3-0, February 27, 2008.

U.S. Department of Defense, *Information Operations Roadmap*, Washington, D.C., October 30, 2003.

———, *National Defense Strategy*, Washington, D.C., June 2008. As of December 23, 2010:
http://purl.access.gpo.gov/GPO/LPS103291

———, *Consolidated Report on Strategic Communication and Information Operations*, submitted to Congress, March 2010a.

———, *Report on Progress Toward Security and Stability in Afghanistan: Report to Congress in Accordance with the 2008 National Defense Authorization Act (Section 1230, Public Law 110-181), as Amended, and United States Plan for Sustaining the Afghanistan National Security Forces: Report to Congress in Accordance with Section 1231 of the National Defense Authorization Act for Fiscal Year 2008 (Public Law 110-181)*, Washington, D.C., April 2010b. As of December 28, 2010:
http://www.defense.gov/pubs/pdfs/Report_Final_SecDef_04_26_10.pdf

U.S. Joint Chiefs of Staff, *Doctrine for Joint Operations*, Washington, D.C., Joint Publication 3-0, September 10, 2001. As of December 23, 2010:
http://purl.access.gpo.gov/GPO/LPS49614

———, *Department of Defense Dictionary of Military and Associated Terms*, Washington, D.C., Joint Publication 1-02, April 12, 2001, as amended through September 30, 2010. As of December 23, 2010:
http://purl.access.gpo.gov/GPO/LPS14106

———, *Doctrine for Joint Psychological Operations*, Washington, D.C., Joint Publication 3-53, September 5, 2003. As of December 22, 2010:
http://purl.access.gpo.gov/GPO/LPS50187

U.S. Marine Corps and U.S. Department of the Army, *Psychological Operations*, Washington, D.C.: Headquarters, Department of the Army, and U.S. Marine Corps, Fleet Marine Force Manual 3-53, Field Manual 33-1, February 15, 1993.

"U.S. Soldiers Reprimanded for Burning Bodies," *CNN.com*, November 26, 2005. As of December 29, 2010:
http://articles.cnn.com/2005-11-26/world/afghan.
us.soldiers_1_jason-kamiya-afghan-soldier-taliban-troops?_s=PM:WORLD

"Villagers Seek Medical Help from ISAF Camp," *Sada-e Azadi Radio*, May 23, 2010. As of January 18, 2011:
http://www.sada-e-azadi.net/Joomla/index.php/en/component/content/
article/1729-villagers-seek-medical-help-from-isaf-camp

"What Is a 'Push' Poll?" American Association for Public Opinion Research, undated web page. As of January 18, 2011:
http://www.aapor.org/What_is_a_Push_Poll_/1485.htm

Yousafzai, Sami, "Trouble: Mistaken for the Mullah," *Newsweek*, October 14, 2002. As of December 22, 2010:
http://www.newsweek.com/2002/10/13/trouble-mistaken-for-the-mullah.html

Za`if, `Abd al-Salam, *My Life with the Taliban*, Alex Strick van Linschoten and Felix Kuehn, eds., New York: Columbia University Press, 2010.

Zucchino, David, "U.S. Fights an Information War in Afghanistan," *Los Angeles Times*, June 11, 2009. As of December 22, 2010:
http://articles.latimes.com/2009/jun/11/world/fg-afghan-information11